COMMUNICATION ACCESS FOR PERSONS WITH HEARING LOSS

Compliance with the Americans with Disabilities Act

Edited by Mark Ross

York Press, Inc.
Baltimore • Toronto • Sydney

Dedication

This book is dedicated to my wife, Helen, for her loyalty,
love, and friendship. Indeed
"eshet chayil, more precious than rubies."

This book was manufactured in the United States of America. Typography by
The Type Shoppe, Inc. Printing and binding by McNaughton & Gunn, Inc..
Cover design by Joseph Dieter, Jr. Book design by Sheila Stoneham.

Supported in part by grant H133E30015 from the U.S. Department of Education, NIDRR, to the Lexington Center for a Rehabilitation Engineering
Research Center (RERC) on Hearing Enhancement and Assistive Devices.

Library of Congress Cataloging-In-Publication Data
Communication access for persons with hearing loss : compliance with
the Americans with Disabilities Act (ADA) / edited by Mark Ross.
p. cm.
Includes bibliographical references and index.
ISBN 0-912752-35-1
1. Deaf—Means of communication. 2. Telecommunications devices
for the deaf. 3. Communication devices for the disabled. I. Ross,
Mark.
HV2502.C65 1994   94-2594
362.4'283—dc20   CIP

COMMUNICATION ACCESS FOR PERSONS WITH HEARING LOSS

Contents

Contributors

Janet Bailey, President
Sign Language Associates
1010 Wayne Avenue, Suite 420
Silver Spring, MD 20910

Diane L. Castle, Ph.D.
National Technical Institute
for the Deaf
Rochester Institute of
Technology
52 Lomb Memorial Dr.
Rochester, NY 14623

Dorinne S. Davis
Hear You Are, Inc.
4 Musconetcong Ave.
Stanhope, NJ 07874

James J. Dempsey, Ph.D.
Southern CT State University
Department of Communication
Disorders
Davis Hall
New Haven, CT 06515

Alice E. Holmes, Ph.D.
University of Florida
Department of Communication
Disorders
Box 100174
Gainesville, FL 32610

Paula Hendricks
Oval Window Audio
33 Wildflower Court
Nederland, CO 80466

Carl Jensema, Ph.D.
The Conference Center, Inc.
1299 Lamberton Dr., Suite 200
Silver Spring, MD 20902

Norman Lederman
Oval Window Audio
33 Wildflower Court
Nederland, CO 80466

Debra L. Lennox
The Conference Center, Inc.
1299 Lamberton Dr., Suite 200
Silver Spring, MD 20902

Harry Levitt, Ph.D.
Speech and Hearing Sciences
CUNY Graduate School
33 W. 42nd St.
New York, NY 10036

Michael Lieske
Sennheiser Electronics
6 Vista Drive
P.O. Box 987
Old Lyme, CT 06371

Mark Ross, Ph.D.
9 Thomas Dr.
Storrs, CT 06268

Howard E. "Rocky" Stone
Self Help for Hard of Hearing
People
7800 Wisconsin Avenue
Bethesda, MD 20814

Karen Peltz Strauss
  National Center for Law and
    Deafness
  800 Florida Avenue, NE
  Washington, DC 20002

Ross Stuckless, Ph.D.
  National Technical Institute for
    the Deaf
  Rochester Institute of
    Technology

One Lomb Memorial Drive
Rochester, NY 14623

Mitchel B. Turbin, Ph.D.
  225 $^1/_2$ 11th Avenue East
  Seattle, WA 98102

EllaVee Yuzon
  Jade Mountain Communications
  228 S. Proctor Ln.
  Pleasant Grove, UT 84062

Foreword

The Americans with Disabilities Act of 1990 (ADA) mandates communication access for people with hearing loss. Such access is often provided with the help of assistive listening devices and systems; however, both those who benefit most from using such systems and many organizations and institutions that must make them available are unfamiliar with them. Even professional providers of hearing health care may neither understand these devices and systems sufficiently, nor appreciate what they can do. This book is designed to help these individuals and organizations.

The role technology plays in hearing health is widely misunderstood. We expect too much. Hearing aids, cochlear implants, and assistive listening devices (ALDs) cannot restore hearing. Each can improve communication, but only if the user understands the technology and makes it work to the maximum extent possible for his or her particular circumstances.

I know someone with a cochlear implant who was deaf before her implant (110 db loss in both ears). During her rehabilitation, she tested her device under every circumstance important to her. She studied her results in a variety of work situations, in church, in her car (radio), in movie theatres, and in social functions. She learned how to manipulate and control the device and how to evaluate upgrades against her original baseline.

She now uses a telephone with minimal difficulty. She participates in large group meetings without using ALDs or interpreters. She sits in the back at church and understands what is being said. She functions well at work and socially.

Her story is not that of the triumph of technology. It is of intelligent use of technology. There are few people like her today. However, understanding that technology is a tool—not a solution—is growing. Bias toward technology in the hearing health field is diminishing. It must if the ADA is to succeed in providing "effective communication" via communication access.

Two years before the ADA was signed into law, some of us were concerned that advocacy had surged ahead of reality. We could envision having legislative requirements for provision of thousands of devices that would go unused or be underused. We have tried to educate consumers to use them, but progress is slow.

Frequently when assistive listening systems are installed, people complain that the system doesn't work for them; that they would do better with another system. The ADA mandates "effective communication." It urges dialog with the intended recipient of access. To avoid backlash to the ADA, and, more importantly, to provide "effective communication," both providers of access and those using access systems need greater education. It will be a slow, frustrating process, with the potential for contentious litigation. Providing communication access and obtaining effective communication are sure to become the most difficult questions facing the ADA. Education is the best answer, which is one reason this is an important book.

Dr. Ross has assembled experts from all parts of the country to help us understand communication access; what it can and cannot do; what it is we, the users of technology and services, must do; and to give us hope. Among the deaf and hard of hearing populations, hope is more important than ever; but hope must be tempered with reality and acceptance of our responsibilities. John Naisbitt has wisely noted that "whenever new technology is introduced into society, there must be a counterbalancing human response—that is, high touch—or the technology is rejected. The more high tech, the more high touch" (p. 39).

Peter Drucker said more than 10 years ago, "In technology . . . we are entering a period of turbulence, a period of rapid innovation . . . . But a time of turbulence is also one of great opportunity for those who can understand, accept and exploit the new realities. It is above all a time of opportunity for leadership."

Mark Ross assumes this leadership role with his initiative in putting together this important book.

Howard E. (Rocky) Stone

REFERENCE

Naisbitt, J. 1982. *Megatrends*. New York: Warner Books.

Preface

When I received my first hearing aid over forty years ago, that little box represented the full extent of assistive devices available to people with a hearing loss. Nevertheless, I felt fortunate to be the recipient of the latest that "modern" technology could offer me. Unlike the immediate prior generation of hearing aids (consisting of two separate, rather large boxes) my first aid was a "monopak" that fit neatly in my pocket and came with a rather unobtrusive (I felt) wire cord that ran from the aid to the button receiver at my ear. Once I used it, I realized just how much I had been missing. It has transformed my life; without the hearing assistance offered, I could not have completed college and graduate studies, or have led the professional and personal life I fortunately have been able to lead.

For me to have asked for other devices (had they been available) would have seemed presumptuous, not only to others but to myself as well. People with any sort of disabilities were supposed to become humble petitioners to the powers-that-be for any sort of special assistance; neither to be seen nor heard by these powers was the ideal and accepted behavior pattern. Basically, society preferred that people with disabilities not be too visible or too demanding. The concept that society had a certain obligation to its handicapped members would have been considered alien and revolutionary. Moreover, as far as aids for hearing-impaired people were concerned, not only were other kinds of assistive devices not available, it was almost impossible to conceive of them.

However, we live in a different era now. As our society evolved, so has our social consciousness. All of us, whether or not we have a disability, have raised our expectations in all areas where society impinges on our lives. The advances we have seen in technology in recent decades have contributed to these raised expectations. What could not be accomplished in the past could hardly be requested. But now we see what modern technology can offer and how new devices can contribute to the quality of our lives. What were once exceptional luxuries are coming to be seen as necessities. Now they not only are being requested, but demanded. Newly developed devices and systems are graduating from the laboratory entering into our lives at an accelerating rate.

Not only in the laboratory are advances being made. In the last 20 years, we have seen a proliferation of social legislation that has an impact on every aspect of our lives. This book was conceived in part because of one such law, the Americans with Disabilities Act (ADA). People with disabilities are in the historically unique and enviable position of being able to advance on two fronts simultaneously: the technical and the legislative. Scientists and engineers now provide devices whose use the law mandates. The fact that the law encourages the full participation of disabled people in our society has, in turn, stimulated manufacturers to create and market an array of new devices.

There is, however, a rather large fly in this ointment. In order to benefit from these legislative and technical advances, people with hearing losses must acknowledge—to themselves, but particularly to others—that they have a hearing problem. Although a hearing loss, as Rocky Stone has eloquently reminded us for years, is "an invisible condition," it is impossible to employ communication access technology without making the condition very "visible," indeed. For most people with hearing losses, communication access means not only the utilization of personal amplification devices (hearing aids), but also such other systems as an infrared or FM receiver, which clearly signal to the observer that the user has a hearing loss. If a hearing-impaired person denies the presence of a hearing loss, or unrealistically minimizes its impact, all the technological developments reviewed in this book would be irrelevant for that person. The greatest challenge we face regarding communication access is neither technological nor legislative, but societal attitudes toward hearing loss—attitudes that seem to be shared fully by many people with hearing losses, particularly those whose losses have their onset later in life.

Many of us have had the experience of suggesting to a friend or relative that he or she get a hearing aid, only to be informed that it was not necessary because he or she could "hear all right"—this comment being made during the course of a conversation in which shouting was the only way to communicate. It is a frustrating experience for all concerned, much more so for the person with the hearing loss. Everyone probably has his or her favorite reason, or reasons, to see a hearing loss as a stigmatizing condition—some people go to absurd lengths to deny its presence (Ross 1992). The main reason seems to be the association of hearing loss with aging and disability stereotypes. In our youth-oriented culture, aging, or at least the appearance of aging, is an offense that must be denied at all costs. We are reluctant to reveal our ages when we grow older, because the number itself becomes associated with age stereotypes. Many people are reluctant to use a visible hearing prosthesis for the same reason. The pity of this attitude is that although someone with a hearing loss may refuse to use such a prosthesis, he or she cannot disguise the communicative consequences of the hearing loss. The person still misses and misunderstands much everyday conversation; social and cultural activities gradually diminish; and lives become more and more constricted. In trying to deny the reality of a hearing loss because of its association with aging, the person with a hearing loss may cause family and friends to ascribe that person's conversational nonsequiturs and confusion to senility—precisely the association that person may have been trying to avoid by refusing to wear a hearing aid.

I should make a distinction here between deaf people who are part of a community of deaf people, who employ American Sign Language as their primary communication system, and hard of hearing people and late deafened adults, those whose primary affiliation is with the "hearing world." In recent years, the former group has been outspoken and assertive in proclaiming their status to the world; neither denial nor shame characterize their view of their own identities. It is the latter group, particularly those with progressive hearing losses dating from adulthood, who all too often deny to themselves and to others the very real evidence of a hearing loss. It is this group—who represent the vast majority of those with hearing losses—whose attitudes toward hearing loss must be changed if communication access is to become a reality in their lives.

It is a truism in the body politic that a large and active constituency receives more attention from the system than a smaller and torpid group. The ADA, which has an impact on all people with disabilities, was passed in part because the advocates were numerous and vocal; but the changes contemplated by the ADA will not be fully implemented without continued pressure from those who hope to benefit from the law. To offer just one small example: some months ago, I received notice of a lecture that seemed interesting to be held in a local community center. I knew, however, from past experience in that center, that I would have a great deal of difficulty hearing the lecturer. I called and asked if they had an "assistive hearing system" available for hard of hearing people. They didn't. They knew little or nothing about such systems, never had heard about the ADA, and were not about to incur the expense of installing an assistive hearing system. My next steps (which I admit to not taking, because I am prone to the same inertia that bedevils most of us) should have been to make additional calls to the center, contact a consumer group (SHHH) and develop a group effort to have the auditorium comply with the law, to contact my congressman's office, to write letters to the Board of Trustees of the institution, to write letters to the editor of the local newspaper, and so on. I have no doubt that one or more of these steps would have brought about the desired result. As my personal pre-New Year's resolution, I am determined to take one or all of these steps the next time I am insensitively and arbitrarily denied "communication access."

Passivity does not create changes. The people who can benefit most from a legal or technical development are obligated to take the activist role in bringing it about. The more people out there advocating, writing letters, making calls, contacting their representatives at the State and Federal levels and, finally, when necessary, triggering the enforcement provisions in the law, the more likely it is that communication access will become a widespread reality in our time. Those who deny the existence of a hearing loss, who will not request communication access because it will signal that they have a hearing loss, not only cut themselves off from economic, social, and cultural activities, but reduce the likelihood of such access for *all* other people with hearing losses.

For me—personally—this book represents more than just another one of my professional activities; I also have a selfish reason for editing this book. I want to help bring about increased communication access so that I—personally—can benefit from it. For example, I am looking forward to spending a night in a hotel without worrying that there will be a fire alarm that I will not hear; I want to be certain that when someone knocks on the door of the room, a light signal will flash; that the telephone will be compatible with my hearing aid and also include a built-in amplifier; that I can respond to a morning wake-up call through a signaling light; that the television set includes either closed captioning or an assistive listening device or both; and that when I go out on the town that I can attend any theatre, auditorium, or movie theatre, knowing that the acoustic conditions are not going to degrade my comprehension of the performance.

Communication access, in brief, can enhance the quality of life of all people with hearing losses, of whatever degree, and whatever their primary mode of interpersonal communication. People with hearing losses are no longer satisfied with hearing aids, a pad and pencil, the commiseration of society; nor should they be—not when there are so many other possibilities that can permit hearing-impaired people to play a fuller role in our society, possibilities from which we all benefit.

A word about the chapters in the book: conscientious and astute readers will note that some overlap and redundancy exists between chapters. Indeed, in several cases, the same figures are used in different chapters. Because of the nature of the topic, it is impossible to eliminate a certain degree of overlap. For example, it is not possible to discuss Induction Loop (IL) large area listening systems, without referring to the telephone coil, or to cover personal FM systems and large area FM systems without explaining FM signals. The intent is to make each chapter a complete exposition of the topic. Because redundancy is an effective learning technique, I hope that the overlap will foster "accessing" the contents of this book.

REFERENCE
Ross, M. 1992. Why people won't wear hearing aids. *Hearing Rehabilitation Quarterly* 17 (2):8–11.

Chapter • 1

The ADA
The Law, Communication Access, and the Role of Audiologists

Karen Peltz Strauss

The Americans with Disabilities Act (ADA) is landmark legislation that will vastly expand the legal protections for and rights of individuals with disabilities.[1] For the first time in the United States, the ADA promises to offer new opportunities in employment, education, and recreation to individuals who are deaf and hard of hearing who were previously denied these opportunities. However, the promises of the ADA will not be fulfilled overnight. More likely, at least a decade or more will pass before the tenets of the ADA are fully accepted and adopted by mainstream society. As with any new civil rights law, education of the American public about the abilities and rights of this newly protected class of individuals will be the first step in the long process toward full integration. Audiologists can play a major role in this educational process by learning about and disseminating information to deaf and hard of hearing individuals about their right to communication access under the ADA.

The ADA is separated into five titles. Titles I, II, and III require private employers, state and local governmental agencies, and places of public accommodation, respectively, to make their programs and activities accessible to individuals with disabilities. Title IV mandates nationwide relay services, and Title V contains a number of miscellaneous provisions, including one that applies the requirements of the ADA to all members of the United States Congress. This chapter discusses in detail the first four of these titles.

TITLE I

Since July 26, 1992, Title I of the ADA has prohibited private businesses that employ at least 25 individuals from discriminating on the basis of disability.[2] As of July 26, 1994, all employers with at least 15 employees will also be covered under this Title. In addition to employers, Title I covers unions, employment agencies, and joint labor-management committees.

Title I's prohibition against discrimination is explicit: a business covered by the Act may not discriminate against a "qualified individual with a disability," defined as an individual who "with or without reasonable accommodation, can perform the essential functions of the employment position."[3] In determining whether an individual is "qualified" to receive protection under the Act, it must be determined whether that individual can carry out those assignments and responsibilities of the job that are *fundamental*, rather than marginal. Consider, for example, the case of a hair salon seeking a qualified employee. The owner has advertised for an individual with experience in cutting and styling both women's and men's hair. Of course, an individual who is hard of hearing may be qualified to perform these fundamental job functions, although the individual may be unable to answer the salon's telephone to schedule appointments. Should answering the telephone occasionally be listed as a peripheral job function, the ADA directs the employer in this situation to transfer responsibility for that task to another employee.

Job restructuring is just one type of reasonable accommodation that an employer may need to make for a deaf or hard of hearing employee. Generally, the ADA requires employers to make whatever modifications or adjustments—or "reasonable accommodations"—to employment that are necessary to enable the employee to perform the essential job functions. These may include acquiring or modifying equipment, such as TTYs, telephone amplifiers, assistive listening devices, or captioning devices. They may also require job reassignment; for example, where job performance by a hard of hearing employee would improve if the employee were moved to a quieter work environment.

The ADA's protection for deaf and hard of hearing workers extends to virtually every facet of employment. Thus, employees are under the ADA's umbrella of protections even when they first apply for an advertised job. For a deaf applicant, this could mean that the employer or employment agency must provide a sign language interpreter to ensure full communication during the job interview. Alternatively, the employer might be required to modify the manner in which its business administers a written qualification examination to an individual whose first language is American Sign Language (ASL). In this situation, the employer might be required to take steps to ensure that the examination measures the applicant's skills and abilities to perform the essential functions of the job, rather than the applicant's English language skills.

The ADA does not permit employers to require potential employees to undergo hearing tests or other medical examinations prior to an offer of employment. Nor may the employer make inquiries about the existence of an individual's disability prior to offering the individual a job. Rather, under the ADA, employers are permitted only to make inquiries about the employee's abilities to perform the functions of the vacant position with or without reasonable accommodations. Once a job has been offered, the employer *may* require the employee to take a medical or hearing examination, but only if *all* entering employees will be required to take the same test, and if the test results are maintained in separate and confidential employee medical records. Moreover, the employer may not use the test results to screen out people with disabilities unfairly; rather the medical criteria must be job related.

Title I of the ADA makes it unlawful for a private employer to discriminate in all other terms and conditions of employment as well, including, but not limited to training, rates of pay, promotion, transfers, job assignments and descriptions, seniority, leaves of absence, social and recreational programs, and fringe benefits. Employers are charged with reasonably accommodating the needs of deaf and hard of hearing individuals to ensure that these individuals have the same opportunities to benefit from these terms and conditions as do nondisabled employees. However, an employer need not make any accommodation that would impose an "undue hardship" on the employer's business; that is, that would create a "significant difficulty or expense" for the business.[4]

To determine whether a particular accommodation would impose an undue hardship, the employer may consider the nature and cost of the accommodation, the overall resources of the business, the number of persons employed, the type of business operation, and the effect of the accommodation on that business.[5] However, the employer may not assert an undue hardship merely because the cost of a particular accommodation is high as compared to the employee's salary. Rather, the ADA focuses on the overall resources of the employer.[6]

Where the resources of a small franchise are concerned, the employer may consider the connection between that franchise and its parent company in determining undue hardship. If the franchise receives considerable funds for its operations from the parent company, then the financial resources of the parent company may be included in the consideration of whether the franchise would suffer an undue hardship if required to provide the accommodation requested. If, on the other hand, the relationship between the two entities consists of a franchise fee only, the employer may consider only the resources of the local franchise. For example, although an independently financed and operated computer store may allege an undue hardship if it is requested to provide an oral interpreter at weekly staff meetings, a computer retail outlet whose finances are closely tied to its parent company may not be able to use that defense.

Deaf and hard of hearing individuals who believe they have been subject to employment discrimination have the right to file a complaint with the United States Equal Employment Opportunity Commission (EEOC).[7] Individuals also have the right, under the ADA, to appeal the EEOC's decision in Federal court. Remedies available to such individuals include back pay, reinstatement, injunctive relief, and attorneys' fees. Injunctive relief might take the form, for example, of an order directing the employer to discontinue a discriminatory condition of employment, or an order requiring the employer to provide a reasonable accommodation that would enable the employee to perform his or her job responsibilities. If it can be shown that the employer has not acted in good faith and has engaged in intentional discrimination, individuals are also entitled to receive compensatory and punitive damages.[8]

TITLE II

Title II of the ADA prohibits state and local governments, known collectively under this Title as "public entities," from discriminating against qualified individuals with disabilities in their services, programs, and activities.[9] Prior to passage of the ADA, Section 504 of the Rehabilitation Act of 1973 already prohibited discrimination in state and local programs that received Federal financial assistance.[10] Title II goes a considerable step further to cover all local governmental services including those that do not receive Federal aid.

Title II applies to virtually any program or service operated by a local government or its instrumentalities. This includes state and local courts, local legislatures and executive agencies, social service agencies, school systems, motor vehicle departments, prisons, public hospitals, libraries, state-operated airports, and transportation agencies. Title II's prohibitions also extend to such private contractors of public entities as state parks, inns, concessions stands, and private, nonprofit counseling agencies.[11]

As in Title I, Title II prohibits public entities from discriminating against deaf and hard of hearing individuals in their employment practices.[12] Unlike Title I, this requirement extends to *all* segments of public entities, regardless of their size.

Under Title II, each public entity must ensure that communication with its applicants and participants who have hearing loss is as effective as its communications with other members of the public.[13] Toward this end, public entities must furnish those auxiliary aids and services that are necessary to afford individuals with hearing loss equal opportunities to participate in and to receive the benefits of the public entities' programs and services.[14] In its regulations implementing Title II, the Department of Justice (DOJ) offers a comprehensive list of these aids and services:

> [q]ualified interpreters, notetakers, transcription services, written materials, telephone handset amplifiers, assistive listening devices, assistive listening systems, telephones compatible with hearing aids, closed caption decoders, open and closed captioning, telecommunications devices for deaf persons (TDDs), [and] videotext displays. . . .[15]

In addition, any other device or service necessary to make spoken or aural information accessible would be considered an auxiliary aid. Thus, a city library may be required to provide assistive listening devices for its weekly children's programs. A city legislature may be required to caption legislative sessions it broadcasts over a local cable television network. A public school may be required to provide interpreting services for deaf parents who want to attend their hearing child's performance in a school play.

To determine the proper auxiliary aid or service, a public entity must give primary consideration to the specific request of the individual with the disability.[16] The governmental entity must honor that choice unless it can demonstrate that an alternate means of providing effective communication exists. DOJ explains that deference to the choice of the individual is "desirable because of the range of disabilities, the variety of auxiliary aids and services, and different circumstances requiring effective communication."[17]

In the analysis to its regulations, DOJ also offers specified guidance on circumstances in which a public entity may be required to provide a qualified interpreter.[18] The analysis explains that, although in some instances, notepads or written materials may be enough to provide effective communication, qualified interpreter services may be needed when the information being exchanged is lengthy or complex. Factors to consider in determining whether the ADA requires the provision of an interpreter, according to DOJ, include "the context in which the communication is taking place, the number of people involved, and the importance of the communication."[19]

Title II of the ADA also places special requirements on state and local governments to make their telephone services accessible to deaf individuals. To some degree, these public entities may turn to the relay services required by Title IV of the ADA to fulfill this requirement. However, the Department of Justice encourages city halls and other entities with extensive telephonic contact with the public to have TTYs on site to ensure direct and immediate access for TTY users. Additionally, "[w]here the provision of telephone service is a major function of the entity, [TTYs] should be available."[20] Finally, state and local governments that offer emergency telephone services, including 911, fire, ambulance, and police emergency telephone services, must provide direct TTY access to those services. The Department of Justice has made clear that, because of the added time needed to relay a telephone call, local governments cannot depend on relay services to handle emergency calls in their communities.[21]

State and local governments are prohibited from assessing charges for the provision or auxiliary aids or services on the individuals requesting those aids or services.[22] For example, courts may not assess the costs of providing computer-aided transcription services in a courtroom on a witness who needs those services in order to testify. Similarly, a city museum that offers tours on audiotape free of charge cannot assess the costs of a written transcript on individuals who cannot hear.

The ADA does not require state and local governments to provide auxiliary aids and services that would result in an "undue financial or administrative burden" on the government or a "fundamental alteration" in the nature of the goods or services it provides.[23] The definition of "undue burden" is virtually identical to the definition of "undue hardship" contained in Title I. Again, factors to consider in determining whether an aid or service will impose an undue burden include: the cost of the aid or service, the overall financial resources available for use in the funding and operation of the public entity's program, the effect of the cost on the resources and operation of that program, and the difficulty of locating or providing the aid or services.[24]

In addition to having to provide auxiliary aids and services to achieve effective communication, each public entity must remove structural communication barriers from its programs and services.[25] One such structural change might be the installation of public pay TTY telephones or permanent smoke alarms at the city hall. Another architectural change may require equipping the auditorium of a state college with a permanent assistive listening system. Similarly, constructing a wall in a state-operated half-way house may be required in order to make an area of a building less noisy for counseling of its hard of hearing residents. Under Title II, it is important to note that each public entity is not required to make *all* its existing facilities structurally accessible, as long as the entity's service or program," when viewed in its entirety, is readily accessible to and usable by individuals with disabilities."[26] Thus, for example, a state that typically administers an examination for teacher certification in a building that is not fully accessible to individuals with disabilities may not be required to make that building accessible, as long as it offers an alternative site for testing these individuals. Where structural changes to an existing facility *are* needed in order to bring the public entity into compliance with the ADA (for instance, in the example above where no alternative site for testing can be located), the Department of Justice's regulations require that those changes be made by January 26, 1995.[27]

All public buildings that were newly constructed or altered after January 26, 1992 are also required to be readily accessible to, and usable by individuals with disabilities.[28] In December 1992, the Architectural and Transportation Barriers Compliance Board issued proposed accessibility standards for new and altered construction.[29] These proposals set forth specific guidelines for the provision of assistive listening systems, public pay TTY telephones, visual public address systems, and smoke alarms for judicial, legislative, regulatory, correctional, and other facilities operated by public entities. The Board will make these rules final after receiving comments from individuals, consumer groups, and businesses affected by the proposals.

Individuals may enforce their rights under Title II by bringing a lawsuit in Federal court or by filing an administrative complaint with the Department of Justice.[30] Additionally, administrative complaints may be filed with any federal agency that provides financial assistance to the program against which the complaint is brought, or with one of eight federal agencies that have been assigned authority for enforcement over specific subject areas.[31] Administrative complaints must be filed within 180 days after the time of the alleged discriminatory act.

TITLE III

Title III of the ADA, which became effective on January 26, 1993, prohibits discrimination on the basis of disability in places of public accommodation.[32] A public accommodation is defined as a private entity, whose operations affect commerce and fall within one of twelve categories:

1. places of lodging, such as hotels or motels;
2. establishments serving food or drink;
3. places of exhibition or entertainment, such as theatres or movie houses;
4. places of public gathering, such as convention centers or lecture halls;

5. sales or rental establishments;
6. service establishments, such as lawyer or doctor offices, dry-cleaners, and gas stations;
7. public transportation stations;
8. places of public display, such as museums or libraries;
9. places of recreation, such as parks or zoos;
10. places of private education;
11. social service center establishments, such as homeless shelters or day care centers;
12. places of exercise or recreation, such as bowling alleys or health spas.

Exempt from Title III requirements are private clubs and entities that are operated and controlled by religious organizations.

As is true for local governments under Title II, places of public accommodation must provide auxiliary aids and services where necessary to ensure effective communication with deaf and hard of hearing people.[33] The list of possible aids and services for entities covered under Title III is identical to that provided by the Department of Justice for Title II. For example, museums that offer aural presentations of their exhibits must make these accessible via assistive listening devices if so requested. Additionally, specific requirements for captioning in this title instruct hotels that provide television viewing in five or more guest rooms and hospitals that offer televisions to provide a means for decoding captions upon request. Although not specifically detailed in DOJ's analysis, it would be this section of the ADA as well, under which a hard of hearing person could request an assistive listening device so that he or she could listen to the audio content of television programs in his or her hotel room. It is this section, also, that directs lawyers, physicians, and other professionals to provide sign language interpreter services for clients, when the exchange of information with those clients is lengthy or complex.

An entity covered by Title III is encouraged to consult with an individual who is deaf or hard of hearing before choosing a particular auxiliary aid or service for that person.[34] Unlike Title II, however, DOJ's discussion of Title III does not instruct places of public accommodation to give *primary consideration* to the choice of the individual requesting the auxiliary aid or service. Although places of public accommodation are not always required to give deference to that choice, such entities do remain under an overriding mandate to ensure effective communication. Thus, a private school would not be meeting its duty to offer effective communication if it insisted on using a notetaker rather than an interpreter for a course that involved active and frequent student participation. Similarly, a court would be violating the ADA's requirement for effective communication if it insisted on providing interpreter services rather than computer-aided transcription for a late-deafened adult who had no knowledge of sign language.

Title III again offers considerable guidance regarding the obligation of places of public accommodation to provide sign language interpreters. Such areas as "health, legal matters, and finances," DOJ explains, are often sufficiently lengthy or complex to require an interpreter to achieve effective communication.[35] Conversely, DOJ offers the example of an individual shopping in a bookstore as one where the exchange of information with the assistance of a notepad and pen, rather than an interpreter, would offer effective communication.[36]

In the past, deaf individuals often were asked to use their children or other family members as interpreters. DOJ frowns upon this practice, and the analysis to its ADA rules makes it clear that even when family members and friends may be qualified to interpret—frequently they are not—their use as interpreters should be discouraged because of questions of confidentiality and emotional involvement.[37]

Title III also sets forth minimum requirements for telephone accessibility. Generally, relay services may be used to enable TTY users to contact places of public accommodations for routine appointments, reservations, and inquiries. Title III does, however, require each place of public accommodation to provide a TTY upon request for *outgoing* calls if the public accommodation typically offers its other customers and clients the opportunity to make outgoing telephone calls on "more than an incidental convenience basis."[38] Thus, although a shoe store may not be required to offer a TTY for outgoing calls, hotels and hospitals fall squarely within this mandate. Additionally, when entry to a place of public accommodation is through a security telephone, a means of making that telephone accessible must be available.

As is true for public entities, the ADA does not require a place of public accommodation to provide an auxiliary aid or service if it can demonstrate that provision of that aid or service would cause an undue burden or would fundamentally alter the nature of its programs or services.[39] Factors to be considered in an undue burden determination are again identical to those listed for entities covered by Title II.[40] However, if the public accommodation can demonstrate that a fundamental alteration or an undue burden exists, it must be prepared to provide an alternative auxiliary aid, where one exists.[41]

Often, more than one public accommodation will be obligated to provide the same auxiliary aid or service under Title III. Two examples illustrate this point. Frequently, troupes of performing artists rent space from local theatre houses for their performances. When performing for the public, both the performing artists and the theatre houses can be considered places of public accommodation under Title III. Similarly, trade organizations, such as medical and other professional associations, frequently gather for conferences in large hotel convention centers. Because in this instance the associations are leasing space, both they and the hotels are again covered under the ADA. In these situations and others such as these, the responsibility to provide effective communications falls on both entities, but may be determined during the lease negotiations between the lessor and the lessee.

The ADA also requires places of public accommodation to make structural changes in the existing facilities to remove communication barriers, where such changes are "readily achievable."[42] A "readily achievable" change is defined as one "easily accomplishable and able to be carried out without much difficulty or expense." For deaf and hard of hearing persons, removal of structural barriers can be achieved, for example, by the installation of flashing alarm systems, pay phone TTYs, permanent signs, and adequate sound buffers. In contrast to older facilities, places of public accommodation and commercial facilities[43] that were constructed for first occupancy after January 26, 1993, or to which alterations were made after January 26, 1992, must be constructed in a manner that is "readily accessible and usable by individuals with disabilities," regardless of whether the structural changes needed are "readily achievable" or are difficult to accomplish.[44] For these facilities, a document produced by the Architectural and Transportation Barriers Compliance Board titled ADA Accessibility Guidelines, or ADAAG, sets forth detailed scoping requirements on accessibility. Although the full panoply of ADAAG requirements affecting deaf and hard of hearing individuals are too numerous to cover here, a few are offered to depict their scope.[45]

Under the ADAAG requirements, newly constructed convention centers, covered malls, and stadiums are obligated to provide at least one interior public pay TTY if these facilities also offer interior public pay telephones. Other ADAAG regulations require the installation of public pay TTYs in banks of four or more telephones of all newly constructed facilities. Similarly, all newly constructed and renovated areas of public accommodation must provide at least one hearing aid compatible pay telephone for each of its floors with one or more single public telephones. ADAAG also directs that at least eight percent of the first 100 rooms in newly constructed or renovated hotel facilities be accessible to deaf and hard of hearing individuals. This mandate directs hotels to provide built in visual smoke alarms, visual notification devices, volume control telephones, and accessible outlets for TTYs near hotel room telephones. Finally, under ADAAG, newly constructed fixed-seating assembly areas that accommodate 50 or more people or that have audio-amplification systems must have permanently installed assistive listening systems.

Individuals who want to enforce their rights under Title III of the

ADA may file administrative complaints with the Department of Justice.[46] If DOJ believes the case is one of public importance or that there is a pattern or practice of discrimination, it may bring the case to Federal court and seek money damages and civil penalties.[47] Individuals, too, may file private lawsuits, but they will only be entitled to injunctive relief (judicial orders directing the entity to cease engaging in discrimination) and attorneys' fees.[48]

TITLE IV

Title IV of the ADA requires all telephone companies to provide both local and long distance telecommunications relay services (TRS) for calls made between users of TTYs and users of voice telephones.[49] The obligation to provide these services becomes effective on July 26, 1993.

A TTY is a machine used to send and receive written signals over telephone lines to other TTYs. The messages sent on a TTY are received on paper or a visual display terminal by the party called. Without a relay service, a TTY user can only communicate over the telephone with other TTY or certain computer users. A relay service bridges the gap between TTY users and voice telephone users by allowing these individuals to carry on near-simultaneous conversations.

To use a relay service, the TTY user places a call to the relay service. The call is then answered by an individual known as a communication assistant (CA) who, in turn, places the call to the hearing party. The CA than converts all TTY messages to the caller into voice and all voice messages from the called party into typed text for the TTY user, until the call is completed. The same process can be performed in reverse, where the call is initiated by a hearing person to a TTY user.

The Federal Communication Commission (FCC) has set forth a stringent list of minimum guidelines regulating Title IV to ensure that telecommunications relay services provided under this title are "functionally equivalent" to telephone services available to conventional telephone users.[50] These rules require, first and foremost, that relay services be provided 24 hours per day, seven days per week,[51] and must be accessible to both the Baudot and the ASCII formats.[52] Moreover, rates for any relayed call can be no greater than the rates charged for a direct dial call from and to the same points of origin and destination.[53]

Relay systems must be capable of relaying any kind of call provided by common carriers.[54] Relay providers may not impose limits on the length, number, or content of calls, and must accept and relay single or sequential calls upon request.[55] CAs are prohibited from disclosing the content of any relayed conversation; all records of conversations must be destroyed upon completion of the relayed call.[56] The

FCC has explained that CAs must act as "transparent conduits relaying conversations without censorship or monitoring functions."[57] Just as a hearing person can use the telephone to communicate any message without limitation, so too, did Congress make clear in the ADA that relay users have this right. Under the FCC's regulations, CAs are also not permitted to alter any part of a relayed conversation intentionally.[58] Among other things, this means that CAs may not interject any opinions or comments into a relayed conversation. The reason for requiring CAs to relay messages verbatim is obvious. Individuals on both ends of a telephone conversation must have a 100% assurance that what they are saying is being accurately conveyed to the directed party. Mistakes made in relaying communications regarding business transactions, medical reports, or airline schedules, for example, could have significant, and detrimental, consequences for relay consumers.

The ADA also requires that relay providers transmit relayed conversations simultaneously, or in "real time."[59] Although there may be some minimal lag time in relaying messages, this requirement ensures that messages are relayed as soon as they are received. Finally, the FCC's rule requires that relay users be given their choice of long distance telephone company when relaying calls, to the same extent that voice callers have this choice.[60]

The quality of services provided by a relay center in large part rests on the abilities of the CAs providing those services. The FCC recognized this, and issued a rule that makes relay providers responsible for ensuring that:

> CAs be sufficiently trained to effectively meet the specialized communications needs of individuals with hearing and speech disabilities; and that CAs have competent skills in typing, grammar, spelling, interpretation of typewritten [American Sign Language], and familiarity with hearing and speech disability cultures, languages and etiquette.[61]

The goal of providing relay services is to close the telecommunications gap between TTY users and voice telephone users. However, the only way this gap can begin to be narrowed is to educate individuals throughout the country about the existence and functions of a relay system. Currently, most hearing individuals are unaware of the functions of relay systems, having had no contact with the services these systems offer. As a consequence, they may be reluctant to use relay services, in part because they may be uncertain about the dependability of such or unaware of their general promises of confidentiality. Moreover, there are many hard of hearing people and others who might need relay services, but who might not be acquainted with these and other services available to the deaf community. In order to educate all these populations, the Federal Communications Commission (FCC) rules require telephone companies to publish information about their relay systems in telephone directories and periodic billing inserts and to make this information available through directory assistance services.[62] Additionally, audiologists can play a role in educating the community about these services by instructing clients who would benefit from using relay services about their existence.

Individuals dissatisfied with the relay services in their state may file a complaint with the local state agency charged with enforcing relay services,[63] or they may be able to file a lawsuit in Federal District Court against the offending carriers. If the state fails to resolve the complaint by final action within a specified amount of time, the FCC is required to receive and investigate the complaint.[64] If, upon investigation of a complaint or the FCCs own initiative, the FCC finds that a carrier has violated Title IV, it can order the carrier to come into compliance immediately,[65] and to pay damages to the complaining party.[66] If necessary, the FCC (or any person affected by noncompliance with an FCC order) can petition a Federal District Court for enforcement of the order by judicial process. Finally willful violations of the Communications Act, including its requirements for relay services, may be subject to criminal penalties, including fines of up to $10,000 and imprisonment.[67]

Title IV also requires all television public service announcements that are produced or funded by the federal government to include closed captioning. This will enable vital information often included in these announcements to finally reach the deaf and hard of hearing public that now relies on closed captions to receive the verbal content of television.

ROLE OF AUDIOLOGISTS IN IMPLEMENTING THE ADA

As professionals in continual contact with individuals protected by the ADA's provisions, audiologists will be affected by the ADA in several ways. First, because they offer professional services, the offices of audiologists themselves are considered places of public accommodation and are covered by Title III of the ADA. Thus, audiologists have an obligation under the Act to ensure that their own services are accessible to all individuals with disabilities. The detailed discussions above of Titles II and III list many of the auxiliary aids and services that audiologists may be required to offer in order to ensure effective communication with their own deaf or hard of hearing clients. In addition, audiologists should be aware that they may need to provide methods of making visually delivered materials available to their clients who are visually impaired. For example, they may need to offer materials in Braille, large-print materials, or taped texts.[68]

For mobility-impaired clients, the existing offices of audiologists will require structural accessibility. As will other businesses, audiologists will have to determine which structural changes can be achieved readily, given the size and resources of their businesses. Similarly, an audiologist who chooses to renovate his or her offices, or construct a new facility, must be aware of the requirement to make such facilities fully accessible to all individuals with disabilities.

Second, audiologists can play a major role in helping their clients to determine the protections to which they are entitled under the ADA. Just as an audiologist may equip a hard of hearing client with a hearing aid to enhance his or her ability to hear over the telephone, so too, should the audiologist equip the client with knowledge about his or her right to hearing aid-compatible telephones in various locations. Similarly, the audiologist who counsels an adult with a recently experienced substantial hearing loss may want to educate that adult about some of the ways in which the ADA can facilitate his or her communication access in various situations. For example, the adult might express an interest in taking college classes, but raise concerns that his or her hearing loss would prevent him or her from achieving success. In this situation, the audiologist should have an elementary knowledge of the individual's basic rights to auxiliary aids and services in publicly funded (Title II) or privately operated (Title III) universities. Equipped with knowledge of the individual's communication needs on the one hand, and his or her basic ADA rights on the other, the audiologist would be in the unique position of being able to counsel the client about ways to continue to participate effectively in the mainstream of society after experiencing a hearing loss.

Third, just as the audiologist can educate ADA consumers about ADA protections, so too will the audiologist be called upon to share his or her knowledge about the ways in which entities covered by the ADA can come into compliance with the Act. The vast majority of employers, public entities, and places of public accommodations have had little, if any, experience attempting to accommodate the communication needs of individuals who have lost their hearing. Conference centers, educational institutions, retail establishments, and other entities are now confronted with modifying their programs and acquiring equipment about which they have virtually no knowledge. Audiologists can play a central role in assisting these facilities to determine appropriate and innovative ways to provide communication access to their clients, customers, and participants. Examples once again illustrate this point. The audiologist may need to counsel a school requesting information on wireless assistive listening systems for its various classrooms. Similarly, a concert hall may need to draw on the expertise of an audiologist to discuss the benefits and disadvantages of

installing a hardwired system in its main theatre. Emplo
turn to audiologists for assistance on how best to accon
offices to meet the communication needs of their ha:
employees. Wherever the demand for communication acc
audiologists should be equipped with sufficient knowledge about the
extent to which the ADA guarantees that access so that they can assist in
offering practical solutions to meet communication needs.

NOTES

1. Public Law 101-336, 42 U.S.C. §12101, *et. seq.*
2. 42 U.S.C. §12111.
3. 42 U.S.C. §12111(8).
4. 42 U.S.C. §12112(B) (5) (A).
5. 42 U.S.C. §12111(10).
6. EEOC Technical Assistance Manual ¶3.9.
7. 42 U.S.C. §12117.
8. Civil Rights Act of 1991.
9. 42 U.S.C. §12131.
10. 29 U.S.C. §794, as amended.
11. 28 C.F.R. §35.130(b).
12. 28 C.F.R. §35.140.
13. 28 C.F.R. §35.160(a).
14. 28 C.F.R. §160(b) (1).
15. 28 C.F.R. §36.303(b) (1).
16. 28 C.F.R. §35.160(b) (2).
17. 56 Fed. Reg. at 35712 (July 26, 1991).
18. The term *qualified interpreter* is defined in DOJ's regulation as " . . . an interpreter who is able to interpret effectively, accurately and impartially both receptively and expressively, using any necessary specialized vocabulary." 28 C.F.R. §36.104.
19. 56 Fed. Reg. at 35712 (July 26, 1991).
20. 56 Fed. Reg. at 35712.
21. 56 Fed. Reg. at 35712, to be codified at 28 C.F.R. §35.162.
22. 28 C.F.R. §35.130(f).
23. 28 C.F.R. §35.154.
24. 28 C.F.R. §35.164; *See also* 55 Fed. Reg. at 35709.
25. This, too, is subject to the undue burden and fundamental alteration defenses of the employer. 28 C.F.R. §35.150(a).
26. Id.
27. 28 C.F.R. §35.150(c).
28. 28 C.F.R. §35.151(a) and (b).
29. *Americans with Disabilities Act Accessibility Guidelines for Buildings and Facilities; State and Local Government Facilities*, 57 Fed. Reg. 60612 *et. seq.* (December 21, 1992)
30. 28 C.F.R. §35.170.
31. These agencies are the Department of Agriculture, the Department of Education, the Department of Health and Human Services, the Department of Housing and Urban Development, the Department of Interior, the Department of Justice, the Department of Labor, and the Department of Transportation.

32. 42 U.S.C. §12181.
33. 28 C.F.R. §36.303(c).
34. 56 Fed. Reg. at 35567.
35. 56 Fed. Reg. at 35567.
36. 56 Fed. Reg. at 35566.
37. 56 Fed. Reg. at 35553.
38. 28 C.F.R. §36.303(d).
39. 28 C.F.R. §36.303(a).
40. 56 Fed. Reg. at 35594.
41. 28 C.F.R. §36.303(f).
42. 28 C.F.R. §36.304.
43. The definition of a commercial facility is broader than that of a public accommodation, and encompasses all facilities intended for nonresidential use by a private entity and whose operations affect commerce. The definition includes factories, office buildings, and privately operated airports.
44. 28 C.F.R. §36.401.
45. 28 C.F.R. Part 36, App. A.
46. 28 C.F.R. §36.502(b).
47. 28 C.F.R. §§36.503, 36.504(a).
48. 28 C.F.R. §36.501.
49. 47 U.S.C. §225, *et. seq.*
50. 47 U.S.C. §§225(a) (3); 225(d) (1).
51. 47 U.S.C. §225(d) (1) (C).
52. 47 C.F.R. §64.604(b) (1). Relay providers must also make provisions for providing uninterruptible power in the event of an emergency. 47 C.F.R. §64.604(b) (4).
53. 47 U.S.C. §225(d) (1) (D).
54. 47 C.F.R. §64.604 (a) (3).
55. 47 U.S.C. §225(d) (1) (E).
56. 47 U.S.C. §225(d) (1) (F).
57. *In re: Telecommunications Services for Individuals with Hearing and Speech Disabilities and the Americans with Disabilities Act of 1990,* Report and Order, CC Dkt. No. 90-571 (released July 26, 1991) at 6 n.2.
58. 47 U.S.C. §225(d) (1) (G).
59. 47 C.F.R. §64.604(b) (4).
60. 47 C.F.R. §64.604(b) (3).
61. 47 C.F.R. §64.604(a) (1). Information about and sensitivity to the cultural and linguistic differences of the deaf, hard of hearing, and speech impaired communities is provided best by individuals from these communities. Accordingly, these TTY user communities have strongly urged state commissions and relay providers to involve them in the training of potential relay employees. For the most part, their efforts have been successful.
62. 47 C.F.R. §64.604(c) (2).
63. 47 U.S.C. §225(g).
64. 47 U.S.C. §225(g) (2).
65. 47 U.S.C. §205(a).
66. 47 U.S.C. §209.
67. 47 U.S.C. §§501 *et. seq.*
68. 28 C.F.R. §36.303(b) (2).

Large-Space Listening Systems

Chapter • 2

Induction Loop Assistive Listening Systems

Norman Lederman and Paula Hendricks

INTRODUCTION

The Architectural and Transportation Barriers Compliance Board states that frequency modulation, infrared, and induction loop assistive listening systems are acceptable for meeting the requirements set forth by the Americans with Disabilities Act of 1990 (ADA Fact Sheet 1990). For the benefit of readers who are unfamiliar with the induction loop (IL) system, a brief overview follows.

The conventional IL system consists of a microphone and/or other audio source and a loop amplifier connected to a wire placed around the listening area (figure 1). The "listening area" may be confined to one person as in the case of a small loop worn around the neck (figure 2), or it may be as large as a theatre (Gilmore and Lederman 1990) with seating for hundreds or thousands of people (figure 3). Portable small area loop systems weighing less than three pounds are also available (figure 4).

Audio signals from the sound source are amplified and sent through the loop wire(s), resulting in the creation of an electromagnetic field. This electromagnetic field is accessed by individuals with hearing aids, cochlear implants, or tactile devices equipped with a "T" (telecoil/telephone) switch, or with portable induction receivers. Switching to "T" disconnects a hearing aid's internal microphone and connects in its place the tiny coil or wire, the telecoil, transforming the hearing aid

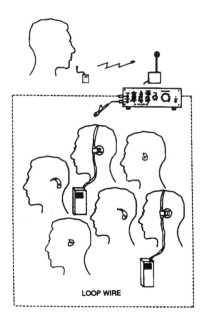

Figure 1.   Typical large-area induction loop systems (IL) can be used by listeners with behind-the-ear and in-the-ear type hearing aids equipped with telecoils. People with wearable induction loop receivers are also shown accessing the system. The speaker's (top) voice is being sent by wireless microphone to the loop amplifier. (Illustration courtesy of Sound and Video Contractor. Artwork done by Oval Window Audio.)

into a *magnetism* sensing, rather than *sound* sensing, instrument. The electromagnetic audio signal from the ILS is picked up by the telecoil, amplified, and directed to the listener's ears. This process of creating an electrical current in a circuit as a result of a nearby current flow is called induction—hence the term "induction loop system."

The end result of an IL system is an amplified reproduction of the original signal, without the signal-degrading effects of room noise and reverberation. As with all assistive listening devices/systems, the IL system enhances audition by *bridging the distance* between the sound source and the listener's ears. The negative effects on signal-to-noise ratios and intelligibility that result from increasing the distance between the sound source and hearing aid microphone has been dramatically demonstrated on a tape recording produced by an IL system manufacturer (ADA and Assistive Listening Systems Demonstration Tape 1992). In this demonstration tape, the output of a hearing aid has been coupled to the microphone of a tape recorder. Recorded in a variety of real-life noisy and reverberant environments, this recording does not attempt to simulate what the hearing-impaired listener *hears*;

Figure 2.   The listener wears an in-the-ear or behind-the-ear hearing aid on "T." The induction neckloop is placed around the neck and plugged into an FM receiver. (Used with permission from Cynthia Compton. 1989. *Assistive Devises: Doorways to Independence.*)

rather, it demonstrates the quality of sound actually being delivered to the person's ears through the hearing aid(s). More relevant to our discussion of distance effects, one excerpt of the tape presents the narrator

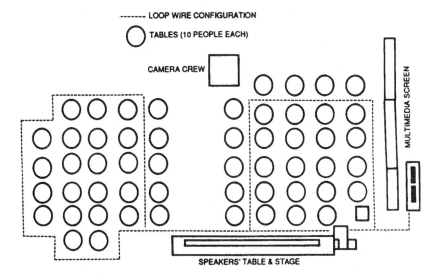

Figure 3.   The Grand Ballroom at the Stouffer Rochester Plaza Hotel measures 134 feet x 66 feet (8,844 sq. ft). One superloop "PRO" and a single satellite induction loop system cover the area. Dotted lines indicate the looped area. Tables outside this area were serviced using the spillover effect. (Illustration courtesy of Sound and Video Contractor. Artwork done by Oval Window Audio.)

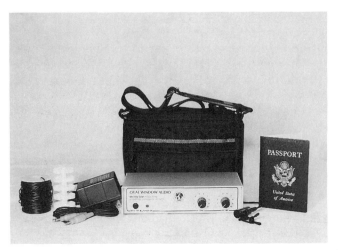

Figure 4. Portable induction loop system. (Illustration courtesy of Oval Window Audio.)

speaking at increasing distances from the hearing aid, beginning at only a few feet away and ending at a distance of twenty-five feet. In the latter part of the excerpt, the narrator's voice is all but lost in a sea of competing room noise. Then, at the same 25 foot distance the hearing aid is switched to "T" and an IL system is used, resulting in clear reception of the narrator's voice. The demonstration clearly substantiates the importance of presenting an input signal with as high a quality as possible to hearing aids and assistive listening systems.

The position of the hearing aid or assistive listening system microphone relative to the sound source is critical. Generally speaking, the closer the microphone is to the source, the better the desired signal relative to unwanted noise will be. In order for hearing-impaired people to have a greater chance of understanding speech clearly, the desired signal should be at least 20 dB above background noise (Gengel 1971; Ross 1972). Naturally, higher signal-to-noise ratios result in a greater likelihood that listener comprehension will occur.

All wireless assistive listening systems in use today have been found to be equally effective when properly installed (Nabelek 1987). Among the most desirable aspects of the IL system is that a telecoil-equipped hearing aid can access the signal, without additional receiver equipment. In other words, the hearing aid user may already be equipped with the receiver portion of an IL system. The potential for universal accessibility, cost effectiveness, low maintenance, efficiency, and unobtrusiveness of the IL system make it attractive for use in schools, meeting rooms, public transportation centers, vehicles, houses of worship, and other public areas.

History of Induction Loop Systems

The electromagnetic induction process is as old as alternating current (AC) electricity and transformer technology. The first portable hearing aid to incorporate a telecoil was the Multitone model VPM in 1938 (Berger 1984), following what may be the first and original patent for the IL system by Joseph Poliakoff of Great Britain in 1937. In fact, the IL system was used in Europe long before it was available in the United States; it made its first U.S. appearance as an imported product shortly after World War II. In the U.S., interest in IL technology reached a peak in the late 1960s and early 1970s. At that time, no standards existed for setting up the systems, nor were standards in place for the manufacture and testing of hearing aid telecoil circuitry necessary for transducing the magnetic field from an IL system. As a result, systems produced by different manufacturers could not be compared, and individual hearing aid response varied. Nevertheless, the consensus of numerous authors during this period seems to indicate that the loop concept was very useful, and afforded a simple, cost-effective method for providing hearing assistance in large areas (Bellefleur 1969; Calvert 1964; Ling 1966; Matkin and Olsen 1970).

Although conventional IL systems have been in use for many years in Europe and are growing in popularity in the U.S., technical limitations have precluded the use of this technology in certain situations. Spillover and inconsistent field uniformity are two limitations that have always plagued IL technology. The term *spillover* refers to the tendency for electromagnetic fields to disregard boundaries and flow out of the immediate looped area potentially resulting in a breach of privacy, and causing confusion for individuals nearby (e.g., adjacent classrooms or offices) who may be switched to "T" using another IL system or telephone.

Sound reception uniformity may also vary due to the predominately vertical orientation of the magnetic field created by a conventional IL system (figure 5). In the case of telecoil-equipped hearing aids worn at ear level, any head motion deviating from the vertical plane, as is common with active children and physically challenged individuals, may result in a partial or almost complete signal drop-out. Remedial approaches have tended to center on redesigning the loop wire configuration, resulting in rather elaborate wiring schemes with partial success at best. As detailed later, only recently have the problems of spillover and inconsistent uniformity been addressed by new techniques and technology.

Twenty-five years ago U.S. manufacturers could not resolve these limitations with the technology of the day and/or did not find it worthwhile to continue in the IL business. The advent of AM and FM

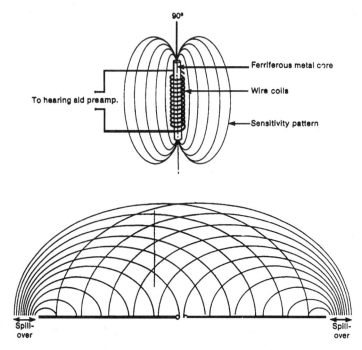

Figure 5. Telecoil's position relative to induction loop. (Used with permission from Cynthia Compton, Assistive Devices Center, Dept. of Audiology, Gallaudet University, 1989.)

"auditory training systems" in the late '60s and early '70s, followed by infrared systems, provided a high level of signal quality and control that had not been attainable with IL technology. At the same time, hearing aid manufacturers were responding to consumer pressure to make their products smaller and smaller, discarding the telecoil along the way. Perhaps the final blow to the hearing aid telecoil was the wave of inexpensive telephones coming onto the market that could not generate the electromagnetic field required by hearing aid telecoils. As a result, the IL system faded into obscurity and it became a cult item assembled by small engineering companies, knowledgeable sound contractors, and technically inclined individuals for consumer groups and personal use. IL technology continued to be popular in Europe, Scandinavia, and the Far East due to its cost-effectiveness and the emphasis that is placed by national health services on including telecoils in hearing aids used in those parts of the world.

Thanks to the efforts of consumer groups, such as the Organization for the Use of the Telephone and Self Help for Hard of Hearing People, change is under way. Further bolstered by international standards for IL systems (IEC 118-4 1981), the Telephone

Compatibility Act of 1988, the Americans with Disabilities Act of 1990, and new advances in technology, IL systems and hearing aid telecoils are experiencing a rebirth in the U.S.

Hearing Aid Telecoils

An examination of IL technology would not be complete without discussing telecoil response, a subject that has been discussed in detail by numerous authors over the years (Gilmore and Lederman 1989; Gladstone 1975; Grimes and Mueller 1991; Rodriguez, Holmes, and Gerhardt 1985; Sung and Hodgson 1971; Sung, Sung, and Hodgson 1974). Until the publication of ANSI S.3.22 (1976) and IEC 118-4 (1981), there were no formal methods for quantifying hearing aid telecoil performance and induction loop system response. A more comprehensive telecoil standard is expected to be published in 1993–94. More detailed and stringent standards have been appended to the IL system standard IEC 118-4 by New York State, manufacturers, and organizations servicing hearing-impaired people (Assistive Devices Advisory Board 1989; IEC 118-4 1981; Specifications: Induction Based Assistive Listening Systems 1991).

Acknowledging that an IL system is only as good as the hearing aid telecoil(s) having access to the system, researchers have evaluated in-situ telecoil responses utilizing hearing aids from a wide range of American and European manufacturers (Grimes and Mueller 1991). Ideally, the frequency response and output characteristics of a hearing aid should not change noticeably when switching from microphone to telecoil mode. Until recently, this ideal was not realized in most hearing aids.

Hearing aid designers are now including active electronics and response-shaping components to telecoil circuitry. As shown in figures 6A and 6B, it is possible to design an in-the-ear hearing aid with a telecoil response that closely mirrors microphone response. It is interesting to note that telecoil-equipped programmable hearing aids that can be controlled by a hand-held remote control device are appearing on the market. With this new technology, not only can tiny hearing aids be switched easily to "T," but the resulting response can be easily altered to meet specific listening situations.

The current ANSI standard calls for the telecoil-equipped hearing aid to be tested full on in a 10 mA/m field. With such a low signal-strength test signal, it has not been possible to obtain meaningful telecoil distortion and noise data due to ambient electrical noise intruding on the hearing aid's performance (see figure 6B). Departing from current test standards, we have obtained useful data by evaluating telecoil response in a 100 mA/m field (with hearing aids set to the refer-

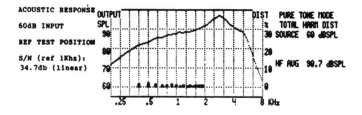

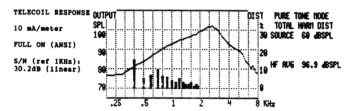

Figure 6A, B. Acoustic versus telecoil response of ITE. Note in 6B the high distortion measurement and noise flow in the test field of 10 mA/meter. (Artwork done by Oval Window Audio.)

ence test position), approximating the "real-life" signals encountered with IL systems and amplified telephones (figure 7). New hearing aid test standards will be based on higher field strengths as well as prescribed telecoil positioning parameters, thereby resulting in more useful specifications.

Now it is possible to build high performance telecoils into all but the tiniest of in-the-canal hearing aids. Manufacturers are attempting to quantify the telecoil output of their products, and the publication of new telecoil standards will ultimately prove to be beneficial for the manufacturer, dispenser, and end user (Cranmer-Briskey 1992).

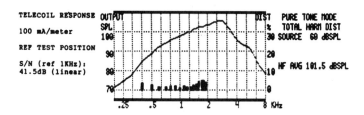

Figure 7. Same ITE from figures 6A, B with 100 mA/meter test field and hearing aid set at reference test position. Note that "distortion" measurement is markedly reduced and meaningful frequency response measurements down to 250 Hz can be made. (Artwork done by Oval Window Audio.)

In 1992, a telecoil forum was presented at the American Speech-Language-Hearing Association National Convention in San Antonio, Texas. For the first time, leaders in the fields of hearing aid technology and assistive listening systems met with dispensers and hard of hearing activists to share information and viewpoints (Cranmer-Briskey 1993). The three-hour meeting resulted in consensus that the telecoil is an invaluable feature for hearing aid users who are aware of its capabilities. It was agreed that it is the dispenser's responsibility to explain and demonstrate to clients the telecoil's potential. The question of whether there should be new legislation requiring telecoils in all hearing aids, as is the case in Europe, was met with a mixed response. Opinions ranged from general resistance to government intervention in such issues, to attitudes strongly favoring legislation. Until new standards are published pertaining to how telecoils are to be positioned and measured in hearing aids, it was generally agreed that the topic of legislation is premature. Panelists shared information that telecoil technology has been greatly improved by smaller packaging, induction loop systems are now available with excellent response and features, and anticipated standards will be more realistic and helpful in clarifying parameters for manufacturers, dispensers, and users. The often ignored hearing aid telecoil is finally being recognized as a potentially invaluable feature that can benefit millions of hearing-impaired people.

NEW DEVELOPMENTS IN INDUCTION LOOP SYSTEMS

Although references are made in this chapter to the problems of signal spillover and inconsistent signal uniformity inherent in conventional IL systems, the use of induction loops remains a viable solution to problems presented by adverse listening situations. Induction loop technology is used in schools, houses of worship, government buildings, offices, and living rooms around the world. If the limitations of conventional IL systems do not present any serious problems to the users (as would occur operating two IL systems simultaneously in adjacent rooms), this form of assistive listening technology continues to be attractive because of its simplicity, low cost, and compatibility with telecoil-equipped hearing aids.

During the period of 1988–90, the U.S. Department of Education's National Institute on Disability and Rehabilitation Research supported the research, development, and commercialization of a new type of induction loop system—the "3-D." The primary objectives of this project were to develop, validate, and commercialize an IL system that eliminated or minimized the problems of signal spillover and inconsistent signal uniformity irrespective of telecoil positioning (Hendricks and Lederman 1991).

The project investigators examined the work that had been done with IL systems. Over the years, systems designers attempted to eliminate or minimize the spillover and signal uniformity problems with elaborate geometric loopwire configurations, shielding methods, and other approaches (Bosman and Joosten 1965). While some innovative approaches may have been successful in a laboratory setting, none was developed in the form of a viable commercially available product that could be easily installed and maintained.

Unlike conventional loop systems that use a continuous, single loop of wire encircling the listening area, the patented 3-D consists of not one, but four loop wires laid out in a prescribed geometric configuration sandwiched into a flexible mat measuring 12 feet x 12 feet and designed to be placed beneath carpeting (figure 8). Multiple mats may be used to cover any size area. Powered by digital signal processing electronics and amplifiers, the end result is a diversity of electromagnetic field orientations confined to the area immediately above the mat(s). The 3-D system (figure 9) consists of the loop mat, VHF wireless microphones, audio mixer, signal processor, and power amplifiers. The modular design results in a flexible, user-friendly system.

The 3-D mat is positioned under the primary listening areas of the room. In a typical installation, carpeting or a rug must be used to cover the 3-D mat (figure 10). The nonskid 3-D mat can be placed alone under the carpet or rug, or the carpet pad may be cut out to accommodate the mat. The mat is then placed into the area formerly taken up by the carpet pad. Once installed, the mat is connected to the 3-D electronics, and signals from the wireless microphones and/or other signal sources are sent through the system. Because the wireless microphones are transmitting on different frequencies to multiple receivers contained in the system's "black box" (rather than to the listeners' individual receivers), multiple wireless microphones may be easily employed.

The 3-D system was field tested in a variety of sites, including: the Governor Baxter School for the Deaf (GBSD) in Falmouth, Maine; the Massachusetts Commission for the Deaf and Hard of Hearing (MCDHH) in Boston, Massachusetts; the Rochester School for the Deaf (RSD) Infant-Toddler Program in Rochester, New York; and Gallaudet University's Assistive Devices Demonstration Room in Washington, DC.

The first objective at the GBSD site was to validate the negligible spillover aspects of the 3-D system. For this reason, two adjacent elementary school classrooms were selected to use the 3-D system. All the students, ranging in age from 5 to 11 years old, with a mean pure-

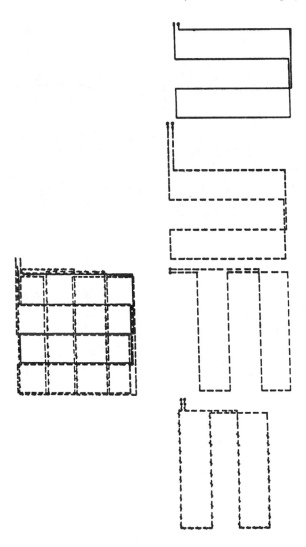

Figure 8.   3-D loop mat wire configuration. (Artwork done by Oval Window Audio.)

tone average in the better ear of 93.5 dB, relinquished their body-worn FM auditory trainers and were fitted with binaural Oticon E39P behind-the-ear aids. These hearing aids were chosen for their power, the sensitivity of their telecoils and for their "MT" switch mode, enabling students to monitor their own voices and nearby sounds, while simultaneously receiving transmissions from the teachers' wireless microphones by way of the 3-D system. During the installation of the GBSD test site, the investigators learned that an interroom attenu-

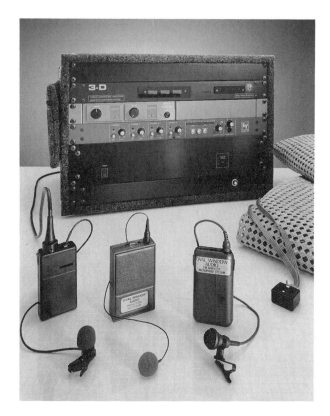

Figure 9. 3-D system with associated hardware and wireless microphones and monitor test receiver. (Photograph courtesy of Oval Window Audio.)

ation of better than −40 dB could be achieved by distancing the mats from common walls by 6 feet. A similar sized conventional IL system exhibited spillover of −13 dB.

A second objective at this site was to evaluate student and teacher responses to using the 3-D system. The students were more enthusiastic and motivated to utilize amplification than ever before. This was attributed to the relatively unobtrusive nature of the students' ear-level hearing aid "receivers." According to the teacher and the consulting audiologist's reports, many students exhibited greater awareness to sound and attended more readily to sounds presented through the 3-D system, ascribable to the overall heightened motivation of the students and the high quality noise-rejecting directional teacher microphone.

From the teachers' perspective the ability to set up and "put away" the 3-D system quickly was an especially attractive feature. Rather than having to check and troubleshoot button receivers and

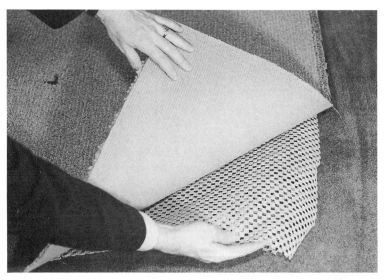

Figure 10.   3-D mat beneath room rug. (Photograph courtesy of Oval Window Audio.)

cords, rechargeable batteries, and other damage prone components of FM auditory trainers, their task was reduced to a basic hearing aid check and test of the teacher microphone.

A 3-D system was next installed at the Rochester School for the Deaf in their Infant-Toddler Program. The primary objective at this site was to test another feature of the 3-D system: the consistent uniformity of the transmitted signals irrespective of hearing aid/head positioning.

As described earlier, all conventional loop systems transmit a predominately vertical field, requiring a vertical orientation of the receiving hearing aid/telecoil for optimum signal reception. Naturally, this was of great importance to RSD. Because infants and toddlers fitted with behind-the-ear hearing aids are in constant motion, their hearing aids depart from vertical most of the time (figure 11). The staff monitored the children's hearing aids and reported excellent uniformity of the signal, regardless of orientation over the mat. Lab tests reveal a maximum +/– 3 dB signal variation with a 360° rotation of the hearing aid situated over a 3-D mat. In comparison, a hearing aid rotated 90° in a conventional loop system exhibits a signal nulling effect greater than –40 dB.

As with the GBSD test site, the positive response from RSD highlighted the 3-D system's ease of use, flexibility, and quality of sound. The option to use multiple wireless microphones was also cited as a definite plus, because a parent or second teacher often was involved

Figure 11. Teacher with wireless microphone and toddler with BTE aid switches to "MT." (Photograph used with permission from the Rochester School for the Deaf and the persons shown.)

with the class. The system's audio mixer accepted a variety of external audio signals from a cassette deck, VCR, and CD player, as well as additional plug-in microphones and wireless microphone receivers. An external mixer output was also available for hooking up to an FM auditory trainer transmitter for use by students who were equipped to receive this signal.

The Massachusetts Commission for the Deaf and Hard of Hearing test site provided a different set of circumstances for hard of hearing adults. The MCDHH's meeting room was equipped with a conventional IL system. The system was helpful to the staff and visitors, but whenever it was in use, hard of hearing staff outside the meeting room could not use their telephones because they picked up spillover transmissions from the meeting room's IL system. A 3-D system was installed in the meeting room to eliminate interference from signal spillover. Staff were finally able to use their telephones, while meetings continued in privacy.

Gallaudet University's Assistive Devices Demonstration Center served as the last 3-D system test site. Here, the 3-D system was demonstrated to visitors and students. Because the Center is also equipped with a conventional IL system, comparisons were made easily between systems. During the field test period, when comparisons were made, most visitors preferred the 3-D system because of its superior signal uniformity. At this site, the overall response to the 3-D system has been positive, and student/visitor awareness of telecoils and IL technology has been greatly expanded.

Dispensers of assistive devices, and demonstration centers such as the one at Gallaudet University, are often called upon by employers and facility managers to recommend a specific system to comply with the Americans with Disabilities Act. By reviewing equipment manufacturer specifications along with the intended application, budget, and installation factors, such professionals can make recommendations for using induction loops, frequency modulation, and/or infrared systems.

As discussed by other authors in this volume, each form of technology brings with it special considerations and pros and cons. Although people may disagree on which type of technology is best for a particular environment, it is generally agreed that there is no *one* perfect system for *every* application. A primary objective of this chapter is to illustrate how contemporary IL technology remains a viable option for providing cost-effective, reliable assistance to hard of hearing people in a variety of settings.

MICROPHONES

All assistive listening systems require a high-quality input signal in order to deliver sound to the listener with the least amount of noise and distortion. There are many different types and styles of microphones available for use with assistive listening systems. Unfortunately there is no *one* "magic" microphone for all situations that will fulfill everyone's listening needs. The more severe the hearing loss, the greater the signal-to-noise ratio (level of desired sound minus the background noise) has to be. This usually means setting up a microphone(s) *close* to the person or persons who are speaking.

The best source of advice on selecting microphones is the manufacturer of the assistive listening system in use. The importance of matching the proper microphone to the system for the specific application cannot be overstated.

NEW APPLICATIONS FOR INDUCTION LOOP SYSTEMS

With advancements in hearing aid and IL technology, new and exciting applications are emerging. In collaboration with Logan International Airport in Boston, Massachusetts, 3-D System investigators were awarded an Innovation Grant from the United States Department of Education in October 1990, to equip two waiting areas of the airport with 3-D systems. These 3-D systems are tied in with the public address systems, thereby providing access by hearing-impaired travelers to announcements, boarding information, and other audio communications. Graphic signs alert the hearing-impaired traveler to the existence of the 3-D systems and provide instructions for using them (figure 12).

An information counter has also been equipped with a 3-D system (figure 13). A hearing aid user seeking information steps up to the booth onto the 3-D mat and switches her or his hearing aid, cochlear implant, or other tactile device to "T." The booth attendant speaks into a directional microphone and the voice message is transmitted through the 3-D mat. The Logan International Airport installations mark the first time wireless assistive listening technology has been made available in a major U.S. public transportation center.

With ongoing support from the U.S. Department of Education, IL

Figure 12. Boston's Logan Airport Terminal C equipped with 3-D systems and signs alerting hearing-impaired travelers to the system's operation. (Illustration courtesy of American Loop Systems.)

Figure 13. Airport information booth equipped with 3-D system. (Illustration courtesy of American Loop Systems.)

system applications are also being explored for passenger and commercial vehicles and other types of transportation.

INDUCTION LOOPS AND ELECTROMAGNETISM

Electromagnetic energy radiates from practically all electronic equipment, as well as from the earth and the sun and all living things. Indeed, it would be impossible to escape all forms of electromagnetism, as suggested by the following excerpt from *The Body Electric* by Becker and Selden (1985).

> Today we're all awash in a sea of energies life has never before experienced, of which the following list of sources only skims the surface:

- Everything that runs on a battery produces a DC magnetic field—from digital watches, cameras, flashlights, and portable radios to car ignition systems.
- Strong magnetic fields are used in industry to refine ore.
- The starting and stopping of an electric train turns the power rail into a giant antenna that radiates . . . waves for over 100 miles.
- Electromagnetic fields vibrating at 60 Hertz (50 Hertz in Europe and other industrialized parts of the world) . . . surround nearly every person on earth from appliances at home and machines at work.
- Over 500,000 miles of high voltage power lines crisscross the United States . . . in effect . . . the largest "radio" transmitters in the world.
- AC magnetic fields . . . emanate from antitheft systems in stores and libraries, and from metal detectors in airports. (pp. 274–75)

Thanks to concerned and responsible researchers, attention has been drawn to the potential hazards of various sources of electromagnetism (Becker 1990; Becker and Selden 1985). But, despite many attempts to unearth the mysterious connection between electromagnetism and latent health threats, researchers are still not certain about the exact characteristics of the electromagnetic fields in question, or the mechanism by which living cells may be altered by their influence. Additionally, no health-related research data specific to IL systems has been published. However, in more than fifty years of use around the world, IL systems have never been cited as a potential health hazard.

We have held informal discussions with scientists and engineers concerning possible health effects of IL systems. The consensus reached is that IL systems generate electromagnetic fields that do not fit the widely published alleged "danger profile" that has been measured around power lines, electric blankets, and other appliances and equipment. Compared with these commonly encountered sources of electromagnetics, IL systems operating to IEC 118-4 standards generate very low power fields that are only present when there is signal modulation. Moreover, unlike most of the aforementioned commonly cited sources of very low-frequency electromagnetic fields, the frequency of the IL field is constantly changing according to the input signal, which typically is speech.

Almost all the equipment, appliances, and power lines cited in research studies generate continuous energy at a specific frequency, typically 60 Hertz. Promising research points to a possible resonance phenomenon that naturally implies a singular constant frequency field interacting with living cells and tissue (Becker 1990; Becker and Selden 1985).

Perhaps the greatest challenge faced by researchers is how to construct studies on human subjects that generate "clean" data not contaminated by the wide range of genetic influences and environmental pollutants present in our lives. Until more is known about the exact nature of electromagnetism and the ways in which it influences living cells, we must weigh the benefits of the electromagnetic-producing devices/equipment with the risks as we know them.

CONCLUSION

This chapter has provided an overview of induction loop assistive listening systems and associated hearing aid telecoils. The IL system offers an attractive option for providing quality assistive listening technology to those who can benefit from it. Practical considerations concerning specific user needs, portability, room design, and budget are but a few aspects that come into play when selecting any type of assistive listening system. The potential for electrical noise interference is very much an issue with all wireless assistive listening technologies, including the IL. It is strongly suggested that a listening check be performed using an induction receiver or telecoil-equipped hearing aid in area(s) being considered for induction type technology. In rare instances, the presence of high-level buzzing or other electrical interference may preclude the use of this type of system. After installation, all assistive listening systems, regardless of type, should be monitored on a regular basis to ensure quality performance. In public spaces signs should be posted alerting hearing-impaired people to the presence and operation of the assistive listening system.

A final point of consideration is the hearing aid itself. After many years of debate, discussion, and published research, hearing aid manufacturers are finally turning their attention to the design of better telecoils in smaller packaging. Knowledgeable hearing aid dispensers and their clients are responding and benefiting. Such consumer groups as Self Help for Hard of Hearing People and the Organization for the Use of the Telephone, have been demanding quality telecoils for years. With the passage of the Telephone Compatibility Act in 1988 and Americans with Disabilities Act in 1990, and with new hearing aid standards under development, IL technology and the venerable telecoil are finally being recognized for their true potential—opening up access to effective communication for millions of hearing aid users around the world.

ACKNOWLEDGMENTS

The authors gratefully acknowledge the cooperation of the participating 3-D system test sites, and the valuable contributions to the validation phase of the project provided by audiologists Eileen Peterson at the Governor Baxter School for the Deaf in Falmouth, Maine; Chris Evans at the Rochester School for the Deaf in Rochester, New York; Cynthia Compton at Gallaudet University in Washington, DC; Robert Gilmore at American Loop Systems in Belmont, Massachusetts, and Douglas Wharton of Spectrum Electronics, Gardiner, Maine, for his technical contributions to the commercialization phase of the 3-D system.

The 3-D system project was supported by an Innovation Grant and Small Business Innovation Research contracts with the U.S. Department of Education.

REFERENCES

ADA Fact Sheet. 1990. Washington, DC: Architectural and Transportation Barriers Compliance Board.

ADA and Assistive Listening Systems Demonstration Tape. 1992. Nederland, CO: Oval Window Audio.

Assistive Devices Advisory Board. 1989. State of New York. Albany, NY.

Becker, R. O. 1990. *Cross Currents.* Los Angeles: Jeremy P. Tarcher, Inc.

Becker, R. O., and Selden, G. 1985. *The Body Electric.* New York: William Morrow.

Bellefleur, P. A. 1969. Critique on current auditory training equipment. *American Annals of the Deaf* 114:790–95.

Berger, K. 1984. *The Hearing Aid: Its Operation and Development.* Livonia, MI: National Hearing Aid Society.

Bosman, D., and Joosten, L. J. M. 1965. A new approach to a space confined magnetic loop induction system. *IEEE Transactions on Audio* AU-13:47–51.

Calvert, D. R. 1964. A comparison of auditory amplifiers in the classroom in a school for the deaf. *Volta Review* 66:544.

Cranmer-Briskey, K. 1992. The ADA spells urgency for telecoil use. *Hearing Instruments* 43(8):8, 12.

Cranmer-Briskey, K. 1993. ASHA Forum: Telecoils, past, present and future. *Hearing Instruments* 44(2):22–27, 40.

Gengel, R. 1971. Acceptable signal to noise ratios for aided speech discrimination by the hearing impaired. *Journal of Auditory Research* 11:219–22.

Gilmore, R., and Lederman, N. 1990. Loop technology: A viable solution. *Sound and Video Contractor* Jan.:70–71.

Gilmore, R., and Lederman, N. 1989. Induction loop assistive listening systems: Back to the Future? *Hearing Instruments* 40(3):14–20.

Gladstone, V. S. 1975. History and status of incompatibility of hearing aids and telephones. *Asha* 17:103–106.

Grimes, A. M., and Mueller, H. G. 1991. Using probe-microphone measures to assess telecoils and ALDs Part I & Part II. *Hearing Journal* 44(6):16, 18–21; Part II 21–24, 29.

Hendricks, P., and Lederman, N. 1991. Development of a three dimensional induction assistive listening system. *Hearing Instruments* 42(9):37–38.

IEC 118-4. 1981. Methods of measurement of electroacoustical characteristics of hearing aids: Magnetic field strength in audio frequency induction loops for hearing aid purposes. *International Electrotechnical Commission Publication 118-4.* Geneva, Switzerland: Author.

Ling, D. 1966. Loop induction for auditory training of deaf children. *Maico Audiological Library Series* 5, Report 2.

Matkin, N. D., and Olsen, W. O. 1970. Response of hearing aids with induction loop amplification systems. *American Annals of the Deaf* 115:73–78.

Nabelek, A. 1987. Comparison of public address systems with assistive listening systems. *Hearing Instruments* 38(2):29–32.

Rodriguez, G. P., Holmes, A. E., and Gerhardt, K. J. 1985. Microphone vs. telecoil performance characteristics. *Hearing Instruments* 36(9):22–24, 57.

Ross, M. 1972. Classroom acoustics and speech intelligibility. *Handbook of Clinical Audiology,* ed. J. Katz. Baltimore: Williams & Wilkins.

Specifications: Induction Based Assistive Listening Systems. 1991. Nederland, CO: Oval Window Audio.

Sung, R. J., and Hodgson, W. R. 1971. Performance of individual hearing aids utilizing microphone and induction coil input. *Journal of Speech and Hearing Research* 14:365–71.

Sung, R. J., Sung, G. S., and Hodgson, W. R. 1974. A comparative study of physical characteristics of hearing aids on microphone and telecoil inputs. *Audiology* 13:78–79.

Chapter • 3

Infrared Systems

Michael Lieske

The transmission of audio signals via infrared (IR) light has been used for well over fifteen years now, making it a time proved technology in assistive listening and translation applications. Audio signals enter infrared systems through the modulator, which superimposes them on a subcarrier; an emitter panel transmits that subcarrier in the form of infrared light, which is picked up by the receivers and turned back into the original audio signal. This process, as well as the different types of equipment that make it possible, is discussed in the greater part of this chapter. The remainder of the chapter deals with different size systems and their applications for assistive listening.

INFRARED TRANSMITTING SYSTEMS

All infrared systems have three basic components: the modulator, the emitter, and the receiver. These components are discussed in the order that they appear in the audio chain.

Modulator

The modulator, also known as the transmitter, is the first component an audio signal encounters as it enters an infrared system. It processes

All photographs in this chapter were supplied courtesy of the Sennheiser Electronic Corporation.

the audio signal so that it can be transmitted via infrared light. This audio processing generally includes limiting and/or companding, pre-emphasis, and some filtering.

The limiting of an audio signal, as its name implies, controls the signal's amplitude (volume). In most modulators, limiters are set up so that they do not affect a standard audio input, but rather, reduce short periods of high volume. Sudden peaks in volume can overload the frequency modulation electronics, causing distortion of the transmitted signal.

Companding circuits, much like limiters, are designed to control amplitude levels in infrared systems. Unlike limiters, however, companders do not limit the dynamic range of a system, but, rather, extend it. These circuits consist of two parts, (1) the compression circuit in the transmitter, which reduces high volume sounds to keep them from distorting the transmitted signal, and raises the level of low-volume sounds to prevent them from being lost in the low-level noise inherent in infrared transmission systems; and (2) the expander in the receiver, which returns all volume levels back to normal. Compressors and expanders are used in many types of wireless equipment to reduce noise picked up during transmission. For example, in a musical production, when the music reaches a thundering crescendo, its volume going into the infrared system is lowered to avoid distorting the signal during transmission. After the signal has been received, the music's volume is again restored to its original level. Similarly, when a lead performer whispers into the microphone, his or her voice is amplified before being transmitted, so that it will not be lost in background hiss. After reception, the voice signal is returned to its lower level and background hiss that might have been picked up is deamplified into the inaudible range.

Pre-emphasis in infrared systems boosts the high-frequency components of the audio signal in order to keep them from being lost during transmission to the receivers. Systems that use pre-emphasis need to have receivers with the ability to reduce or de-emphasize the high-frequency components of the received signal, in order to recreate the original signal.

It is the modulator's second function, from which it derives its name. Besides processing the audio, the modulator also frequency modulates (FM) the audio signal onto a radio frequency (RF) subcarrier. This process (see figure 1) is similar to that used by FM radio stations, except that this FM signal is transmitted via infrared light instead of RF waves. Most infrared systems on the market today use 95 KHz as the RF subcarrier for single channel, or mono-systems and 95 KHz and 250 KHz in two-channel, or stereo-systems. This de facto standard allows the use of any manufacturer's receiver in all locations with installed IR listening systems. These 1- and 2-channel systems com-

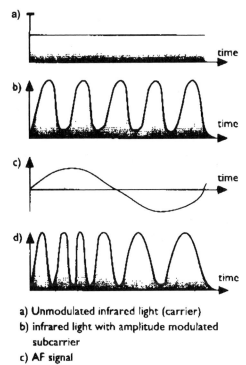

a) Unmodulated infrared light (carrier)
b) infrared light with amplitude modulated
   subcarrier
c) AF signal
d) infrared light modulated with frequency
   modulated subcarrier

Figure 1. FM modulation of a subcarrier in an infrared system: (a) unmodulated infrared light (carrier); (b) infrared light with amplitude modulated subcarrier; (c) AF signal; and (d) infrared light modulated with frequency modulated subcarrier.

monly use wideband modulation to allow for as wide a dynamic
range as possible. To achieve these results, deviations of up to +/–50
KHz are the standard in 1- or 2-channel systems. Multichannel systems, ranging from 3 to 15 channels, employ narrowband modulation
with deviations of +/–7 KHz. This limits the system's dynamic range
and frequency response, making it more suitable for voice communications than for high-fidelity music applications. Because of these limitations, multichannel systems generally are not used to assist hard-of-
hearing persons.

Emitter

Infrared emitters have only one function: to take the RF signal from
the modulator and send it out on rays of infrared light. This is done

by varying the output intensity of infrared emitter diodes with the signal from the modulator. In recent years, much progress has been made to increase both the output and life expectancy of these diodes, allowing manufacturers to build emitter panels that will last longer and cover greater areas than ever before. The area an emitter can cover also depends on the number of diodes it contains. In much the way that one or two lightbulbs are sufficient to illuminate a standard living room, it takes only a few infrared diodes to cover small areas. (Figure 2 shows an emitter capable of covering up to 400 square feet.) To advance this analogy, it would take several hundred lightbulbs to illuminate your neighborhood coliseum, which means that it would also take many more infrared diodes to cover all seats in that same coliseum adequately. Large area panels, such as that in figure 3, use over 100 infrared emitter diodes and can cover up to 11,000 square feet of unobstructed space. Many infrared systems combine the modulator and emitter in one package for ease of use.

Receiver

Infrared receivers reverse the process the audio signal has gone through to this point. Initially the receiver picks up the infrared light with a receiving diode. The RF subcarrier signal is then demodulated

Figure 2.   Small-area transmitter capable of covering 435 square feet.

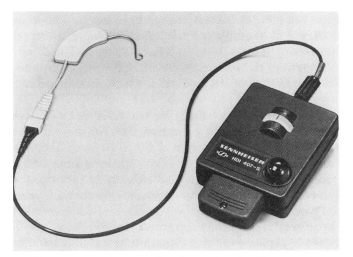

Figure 3.   Large-area emitter capable of covering 11,000 square feet.

and the audio signal is retrieved. If any companding and/or pre-emphasis occurs in the modulator, the audio signal is expanded and de-emphasized before being amplified and sent to the earphones or hearing aid accessory.

There are several basic types of receiver: the body pack, the stethophone, and the headphone receiver. The body pack receiver (figure 4) picks up an infrared signal and then sends it to a pair of head-

Figure 4.   Body pack receiver with silhouette.

phones or to hearing aid accessories plugged into it. The stethophone receiver (figure 5) resembling a medical stethoscope, incorporates the headphones and receiver into one compact unit. The third type of receiver integrates the infrared receiver electronics right into the ear cup with the driver elements, and, thus, it looks like an ordinary headphone.

All types of receivers usually have at least an on/off control, to preserve battery lifetime, and volume control. For those receivers with more than one channel, it is important to have a channel selector switch to enable the user to receive both channels simultaneously for stereo reception, or each channel individually to facilitate the use of these receivers in single channel installations that are common in houses of worship, theatres, nursing homes, and other venues (see figure 6). Most receivers on the market rely upon rechargeable batteries with operating lifetimes ranging from three to seven hours. Use of this type of batteries, especially in large theatre installations, results in saving both money and the environment. Most large installations will give receivers at the ticket booth or especially designed stands for a small deposit, a driver's license, or a credit card to ensure their prompt return.

Receiver Accessories In addition to headphones and earbuds there are several other types of accessories available. These accessories include neckloops, silhouettes, and direct audio input cables. Neck-

Figure 5. Stethophone receiver.

Figure 6. Two channel receiver.

loops, which plug right into the audio output of any body pack receiver or stethoscope receiver equipped with an audio output, are wire loops that use the audio signal supplied by the receiver to create an electromagnetic field. This electromagnetic field is picked up by small coils, called tele- or T-coils—integrated into many hearing aids—and turned back into audio signals. Silhouettes operate in the same manner as neckloops, but on a much smaller scale. They are integrated into small housings that are worn behind the user's ear, next to his or her hearing aid. The third type of hearing aid accessory, the direct audio input cable, connects the receiver directly to the hearing aid(s) via special plugs found on hearing aids equipped with boots. This is probably the best and most reliable way to feed audio directly to a user's hearing aid(s).

INFRARED SYSTEM TYPES AND APPLICATIONS

There are basically three types of infrared assistive listening systems: for small, medium, and large areas. Each of these systems consists of a modulator, emitters, and receivers. All systems have one modulator per channel and are differentiated by the number of emitters and receivers they use.

Small-area Systems

This category commonly uses a modulator/emitter combination enclosed in a package small enough to fit on top of most television sets or stereo systems. The area covered by these systems is usually between 100 to 450 square feet, which is enough to cover the average living room, the most likely location in which these systems will be found. In general, one of these personal listening systems consists of the modulator/emitter and one or two stethophone receivers. Generally, the receivers are powered by rechargeable batteries with a lifetime from three to seven hours, depending on the manufacturer. Figure 7 shows a typical personal listening system, including a small area transmitter with an integrated charger for the receiver's batteries, and the receiver. Systems such as these are usually connected to the audio output or earphone jack of television sets or stereo systems. If there are no usable audio outputs on the equipment, microphones available from most manufacturers can be used to pick up the audio signals directly from the speaker.

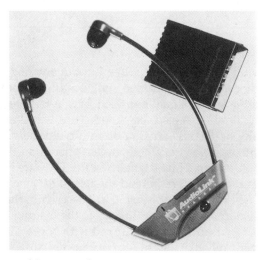

Figure 7.   Personal Listening System.

Medium-area Systems

Medium-area infrared systems (see figure 8) can cover from 500 square feet to several thousand square feet, at which point large-area systems become more cost effective. These systems usually encompass a modulator/emitter combination capable of covering a few hundred square feet, several slave emitter panels, which serve only to increase the coverage area and not affect the audio processing, and receivers. Depending upon the size of the room, slave emitters are added to the system until the entire seating area is covered by infrared light. With more than five or six receivers, it becomes cumbersome to charge all batteries in the transmitter, so manufacturers of infrared systems sell charging strips that will fully recharge all batteries simultaneously overnight. Common applications for medium-area systems are court and conference rooms, small theatres, auditoriums, and houses of worship.

Large-area Systems

These systems are used in venues exceeding 5,000 square feet, such as some large houses of worship, cathedrals, theatres, opera houses, civic centers, and so forth. Large-area systems usually consist of a modulator, several emitter panels, and many receivers. The size of a particular venue determines the number of emitter panels needed. Logistics of installing infrared systems in large rooms are somewhat more involved than those for small and medium-area systems, because large rooms normally contain visual obstructions, such as balconies and columns, which require special consideration in the placement of emitter panels.

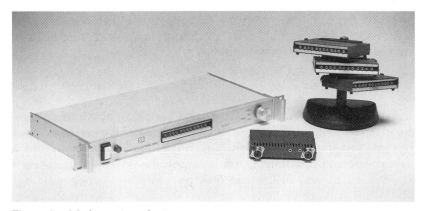

Figure 8.   Medium-Area System.

SET UP AND INSTALLATION OF AN INFRARED SYSTEM

Installation of an infrared system requires only three types of components, as previously mentioned. These are the modulator, the emitter (sometimes combined into one unit), and the receivers. No matter what type of room the system will be installed in, there can never be more than one modulator per channel in the same room, because that would cause severe interference when the receivers pick up multiple signals on the same frequency simultaneously. Depending upon the size of the room, however, there can be multiple emitter panels to ensure adequate coverage of the entire area. The modulator should be placed as close to the audio source as possible, in order to avoid picking up noise on long audio cables. When choosing locations for the emitter panels it is important to consider locations that are not obscured from the audience, because the panels emit light, which cannot travel around corners or through walls and curtains. Although infrared light shares many of the reflective properties of visible light, and can, therefore, reach places not in the direct line of sight of a receiver, the ideal mount for an emitter panel is that where most receivers in the audience can view it clearly. The more direct the line of sight between the emitter panel and the receiver, the higher the signal-to-noise ratio of the received audio signal and the more reliable the reception. Normally, in theatres, emitter panels are located next to or above the stage, where all patrons have a good view of them. It is best to avoid direct sunlight, as it contains a large infrared component that can interfere with infrared reception. However, if the receiving diode is shaded from direct sunlight and the infrared system has been installed properly, reception will not be influenced by ambient light. Receivers must be stored in such cool, dry places as carrying cases or storage drawers, and their batteries must be recharged periodically in order to ensure proper operation when handed out to the audience. A word on hygiene—the receiver's earbuds or ear cushions should be cleaned regularly in mild soap or detergent, after having been removed from the receiver or headphone.

CONCLUSION

Infrared systems are suitable for a great variety of applications ranging from small living rooms to large theatres and concert halls. Their obvious benefits are the extremely high quality of signal transmission, ease of use and installation, and immunity to outside interference. A large installed base of users in theatres, churches, public buildings, and other locations that are compatible with home systems allows the use of privately owned receivers in public places.

Chapter • 4

FM Large-Area Listening Systems
Description and Comparison with Other Such Systems

Mark Ross

In the previous two chapters, infrared and induction loop wide-area listening systems were described. This chapter presents the last of the trilogy of large-area listening systems: FM systems. Secondarily, we compare the relative strengths, weaknesses, and performances of all three large-area listening systems.

It should be noted at the outset that the principal advantage of these systems (and individual systems as well) relates to their ability to "bridge" the distance between the sound source and the listener. The induction system accomplishes this by utilizing an electromagnetic field, the infrared system employs an invisible light beam, and the FM system connects the source and the listener via a frequency modulated (FM) radio wave. All these systems either eliminate or reduce the negative impact of distance, noise, and reverberation on the speech perception abilities of people with hearing losses.

Poor environmental acoustics do not affect hearing-impaired people only. The rationale, however, for the use of any large-area listening system rests upon the *greater* impact of acoustic factors on the speech perception skills of this population. That is, acoustic conditions that may be suitable, or only mildly problematic for nonhearing-impaired people, can have a severe effect on people with hearing losses (Ross 1992). To illustrate this point and the improvement in speech perception that can be obtained with an FM system, I will briefly review a study in which I participated (Bankoski and Ross 1984).

This study evaluated the effects of seating location in an auditorium on the speech perception scores of nonhearing-impaired college students, and compared their performance to that of older hard of hearing adults using hearing aids and two different FM systems. The Tri-Word Test of Intelligibility (TTI) (Sergeant, Atkinson, and Lacroix 1979) was the test instrument used. The advantages of using the three words in succession that make up the TTI are: (1) normal co-articulatory effects are permitted to occur between words; (2) the test is sensitive to the phonetic "smear" (masking) effects produced by reverberation; and (3) the inclusion of a 150-word corpus (rather than the usual 50-item test list) makes for more reliable and valid test results.

In figure 1, the speech perception scores for the nonhearing-impaired control group of college students are shown by seat location in the auditorium. As can be noted, the poorest scores were in the *front* of the auditorium, whereas the best scores were obtained by the students seated in the middle and rear. We ordinarily expect that people hear better when they are closer to the loudspeaker, and, indeed, this is the recommendation that clinical audiologists often make to their clients. In this case, however, the situation was reversed. The explanation is simple: the loudspeakers were located on a raised stage and the acoustic stimuli simply "went over the heads" of the students in the front rows. (Because this auditorium is in daily use at a university, I am tempted to assume that much of the gist of the material presented by the instructors also probably went over the student's heads!)

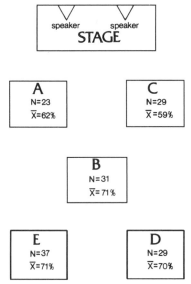

Figure 1. Speech perception scores for nonhearing-impaired college students in a college auditorium.

Even in the best locations, however, scores obtained by the students were much poorer than they should have been. We also tested a subgroup using earphones in the clinic, rather than in the auditorium. The comparisons are shown in figure 2, where it can be seen that the scores obtained in the clinic average some 20% to 25% above those obtained in the auditorium.

Figure 2 also shows the scores for the nonhearing-disabled group plotted by word position and general location in the auditorium. As can be noted, the scores for the second and third words are poorer than those for the first word. This is probably due to the fact that reflected energy from the first word masked subsequent words. If auditorium acoustics give this much difficulty to nonhearing-disabled young college students, we could expect that hearing-impaired people would have a greater problem. This is, indeed, what happens.

Figure 3 compares the differences in speech perception scores between the two groups and demonstrates the improvement produced by both FM systems tested. The filled circles are the scores of the hard of hearing adults, and the open circles are the average scores obtained by the nonhearing-disabled college students in the auditorium. There is

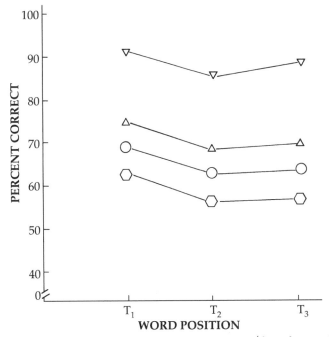

Figure 2. TTI scores for nonhearing-impaired college students by word position and seating location. Key: Headphones ∇; Sections B, D, and E Δ; Sections A, C ○; Overall O.

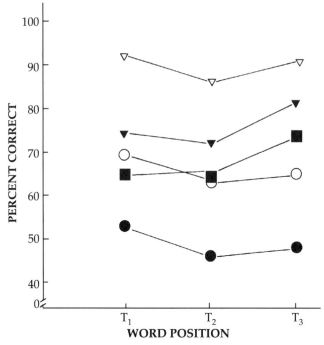

Figure 3.   TTI scores for hearing-impaired and nonhearing-impaired listeners. Key: Headphones, normal hearing ▽; overall scores, nonhearing-impaired O; Usual listening condition, hearing-impaired listeners ● ; first FM system ■; second FM system ▼.

a 15% to 20% difference in scores between the two groups. The filled squares and triangles plot the scores obtained by the hard of hearing adults with the FM systems. Note that *with the FM systems, the scores of these older hard of hearing people were equal to and better than those obtained by the younger, nonhearing-disabled college students.* Note, too, that with the FM systems, the effect of reverberation upon the second and third words was eliminated.

The need for wide-area listening systems for hard of hearing people is graphically supported by the data portrayed in figure 3. Listening to a lecture or performance under such circumstances creates a stressful listening experience—one in which the stress can be minimized given the possibilities inherent in current technology. The listening task in such circumstances should reflect only the limits imposed by a person's residual hearing, not the additional difficulties imposed by the acoustic conditions. Speaking as a hard of hearing person, I can attest to the tremendous difference in speech comprehension (as well as in less fatigue and more enjoyment) a wide-area listening condition can make in this type of listening environment.

THE RADIO FM LINK

The "radio link" eliminates the negative impact on speech perception attributable to poor acoustic conditions. Large-area FM auditory access systems accomplish this in exactly the same way as do individual FM auditory training systems used by hearing-impaired youngsters in classrooms—the one difference being that a more powerful transmitter is required. As in individual systems, the wide-area FM system can be conceptualized as, and basically is, an FM radio transmitter–receiver combination. The desired signal (speech, music, etc.) is "broadcast" to the listener in the audience, who picks up the signal with a special FM radio receiver tuned to the transmitting frequency. A major advantage of a radio wave is that the signal can be transmitted with equal strength throughout the desired listening area; that is, there are no fluctuations in intensity, no "dead " spots, and no positional effects (a listener does not have to face the transmitter or keep the receiver uncovered, as is the case with infrared transmissions). The range of large-space listening systems is over 300 feet, thus making them suitable for almost any listening situation, indoors and out. Installation of remote antennas can increase the effective range

The Federal Communications Commission has allocated a limited band of FM frequencies for use with hearing-impaired people, either in classrooms or in wide-area listening situations. The permitted FM frequency range extends from 72 MHz to 76 MHz. (This can be compared to the commercial range of frequencies extending from 88 MHz to 108 MHz.) Although the commercial frequency band *can* be used in a large-area listening system, such systems would be subject to interference, inappropriate and poor quality receivers, as well as to other problems. Although the cost is higher, there are significant advantages in using only the frequencies and FM auditory access systems specifically designed for hearing-impaired people.

The total permitted range of FM frequencies can be divided into either 40 narrow-band channels or 10 wide-band channels (see table I). Each narrow-band channel is 50,000 KHz wide and each wide-band channel is 200,000 KHz wide. In the FM modulation process (where the audio signal varies—modulates—the frequency of the carrier wave), the wider the bandwidth of the FM channel, the higher the fidelity of the resulting audio signal. Ordinarily, the narrow-band

Table I. Center Frequency of FM Wide-band Auditory Assistance Systems

Channel A:	72.1 MHz	Channel F:	75.5 MHz
Channel B:	72.3 MHz	Channel G:	75.7 MHz
Channel C:	72.5 MHz	Channel H:	75.9 MHz
Channel D:	72.7 MHz	Channel I:	74.7 MHz
Channel E:	72.9 MHz	Channel J:	75.3 MHz

channels are used for classroom instruction, although they may also be used in a large-area system. The narrow-band channels provide audio signals approximately 45 dB above the base level of noise in the transmitter (Boothroyd 1992), which would be perfectly satisfactory for most hard of hearing people. The fidelity of large-area listening systems is usually somewhat greater because they ordinarily employ the wideband channels. For both wide- and narrowband channels, the key consideration is the type and nature of interfering signals received on site, the goal being to select a transmitting–receiving frequency that is subject to the least interference.

FM LARGE-AREA LISTENING SYSTEMS

The advent of the Americans with Disabilities Act (ADA) has encouraged manufacturers who previously focused on manufacturing only FM systems for children in classrooms to provide large-area listening systems also. In addition, the law has stimulated the entrance of other companies into the same market. The market potential of large-area listening systems, because of the stimulus by the ADA, and because of the "graying of America," is simply enormous. By the time this book appears, I believe that more and more companies will be manufacturing and distributing large-area listening systems. Before the end of this century, I expect (and certainly hope) that every auditorium, theatre, lecture hall, or house of worship in the country will include a large-area listening system routinely. The technology is well known and effective; differences between and within types of systems relate more to secondary factors than they do to the fundamental principles of operation (which does not imply that such factors may not be of value for specific individuals or circumstances).

Emphasis in this section is on description of a generic FM wide-area listening system, its operation, and an overview of features available in current systems. Pictorial examples of a number of systems are displayed, and a list of representative companies and their addresses are given in the Appendix. Specific products are *not* recommended: decisions to purchase one such system over another depends upon cost, service, maintenance, and operational factors that are beyond the scope of this chapter. The chapter does include some features prospective purchasers of such systems should consider, particularly those pertaining to the receiver portion of the FM system.

The interconnections portrayed in figure 4 depict a generic installation of a large-area FM auditory access system. The figure assumes an existing public address (PA) system, and includes provisions for external microphones and the alternative coupling arrangements pos-

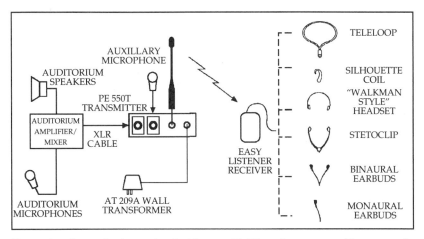

Figure 4.   General schematic of wide-area FM listening system. (Courtesy of Phonic Ear Inc.)

sible with an FM receiver. The specific segments in the figure are discussed in the next several sections.

Wide-Area FM Transmitter

The desired signal is delivered to the transmitter in several ways. In the typical arrangement, as seen in figure 4, input to the transmitter is provided by patching into one of the auxiliary outputs in either the amplifier or mixer (this may be the "tape" or "line" out jack). This connection does not interfere with the normal function of the sound system. It is desirable to orient the antenna vertically and locate it—or a remote antenna if one is being used—toward the front of the listening area and above the seats. It is best to avoid placing an RF antenna on or near structural metal elements (e.g., ductwork, metal, plaster lath, foil-backed insulation). Such elements can absorb radio energy and reduce the operating range of the system.

For locations in which there is no existing sound system, one or more microphones can be plugged into the transmitter directly. It is important that transmitters include the capability to accept microphone inputs, so the device can be used in all types of facilities. It is necessary to ensure that the input level to the transmitter is appropriate—too little will provide an inadequate signal and too much will overload the transmitter and produce distortion. Some systems include an Automatic Gain Control (AGC) unit at the input to control the intensity level of the signal, whereas others permit gain adjustments that can be monitored by a Light Emitting Display (LED) (see figures 5 and 6).

Figure 5A and 5B.   Front and back of wide-area Comtek FM transmitter. (Courtesy of Comtek, Inc.)

Figure 6.   Telex Base Station FM Transmitter. (Courtesy of Telex Communications, Inc.)

Transmitters should permit easy access for changing the operating frequencies. The newer transmitters have the potential to use any one of 10 wide-band FM frequencies, as compared to 8 frequencies in the older units. This may be useful in large facilities where a number of FM auditory assistance devices are being used. Different manufacturers use the same operating frequencies, although they may code them differently. This can permit receivers of one company to be used with a transmitter from a different manufacturer, if necessary. The most common wide-band frequencies seem to be those found in table I. Examples of several representative wide-area FM transmitters can be found in figures 5 and 6.

Murphy's law (anything that *can* go wrong *will*) applies to auditory access systems as much as it does to most other aspects of life, particularly when technology is involved. Some problems will no doubt occur after the initial installation. The problems may relate to the placement of the antenna, to inadequate input signal strength, to interference from other radio services, and so forth. *Do not* get discouraged; such problems can usually be solved. Contact the manufacturer's service representative if local talent cannot solve the problem.

In the interests of flexibility, it may be advisable for locations that utilize one of the large, nonportable transmitters shown in figures 5 and 6 also to make available a wearable transmitter (see Chapter 5). This type of transmitter would be more suitable than a large transmitter in smaller auditoriums or lecture rooms. The same receiver can be used with both a wide-area and a portable FM transmitter provided that it is tuned to the correct transmitting frequency.

FM Receiver

FM receivers in auditory assistance listening systems are merely FM radios tuned to the specific frequency employed by the FM transmitter. Beyond this basic operating principle, however, receivers offer different features that may be valuable, or necessary, in specific circumstances. Considering the rapidity of change in this area, the list of features below may not be complete. One generalization that applies, now and in the future, is that specific circumstances, that is, locations and populations, will help to determine what type of receiver is required. (List order is *not* an indication of a feature's relative importance.)

1. Some receivers are preset to a factory-selected, fixed channel. The channel frequency is designated by a number or a letter. Although this feature reduces receiver flexibility, it may be more convenient to use in situations with only one listening location or with certain populations (e.g., an auditorium in a nursing home).

2. Other receivers have tunable receivers, which can be set to any one of 8 or 10 channels for use in locations where several transmitters are broadcasting on different frequencies (see figure 7). Prospective purchasers must define the conditions under which

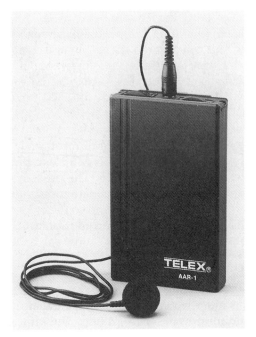

Figure 7. Telex AAR-1 tunable FM receiver. (Courtesy of Telex Communications, Inc.)

the FM wide-area listening system is to be used in order to determine whether a tunable or a fixed-frequency receiver would be suitable.

3. Some receivers employ rechargeable batteries and others use disposable batteries. Some receivers can operate with either a rechargeable or a disposable battery (*do not attempt to recharge a disposable battery*). No one method of powering the receiver is intrinsically superior to another; it all depends on the logistic conditions (e.g., who takes care of checking the batteries and the dependability of this person) in a particular location. At least one battery recharger serves a dual function—it recharges and changes the channel on the receiver (see figure 8).

4. A light that indicates when the receiver is on is desirable. Even more desirable is a light that glows when battery voltage falls below a critical level. This feature is common on individual FM systems, although not on wide-area systems. Because in the average use situation, it is difficult to assure that receivers are properly charged, or are using fresh batteries, such a warning light would be a valuable addition to a large-area FM receiver.

5. An environmental microphone with an adjustable gain control is another desirable feature. If the FM system permits reception of

Figure 8. Large-area Chaparral FM listening system, including transmitter, chargers, and receivers. (Courtesy of Chaparral Communications, Inc.)

the primary source only, it will not allow an occasional remark to or from one's companion in the next seat. For some performances, lectures, and so forth, such communication with a neighbor may not be desirable (at least from the point of view of a lecturer or those in nearby seats), however, it would be convenient to be able to converse with the person in the next seat without removing the earphones. An example of an FM receiver that includes an environmental microphone is shown in figure 9.

Figure 9. FM receiver with environmental microphone. (Courtesy of Williams Sound Corp.)

6. Most wide-area FM receivers incorporate only a minimum of controls, perhaps only an "on–off" switch and a volume control. For many locations, no more complicated system may be required. However, because these receivers are meant to be used by hearing-impaired people, it would be desirable if the receivers included such additional characteristics as individual ear volume, and tone and output controls. Audiologists spend a lot of time ensuring that wearable hearing aids embody desired electroacoustic characteristics for a particular person, based on the premise that such characteristics improve speech comprehension; the same logic applies to receivers in large-area listening systems.

7. All FM receivers must deliver the signal to a person's ears in some fashion. As shown in figure 4, this can be accomplished in a number of ways. Probably, the most convenient for people who do not use hearing aids would be a headset (earphones). However, a number of other coupling options are available, including those designed to be employed with a personal hearing aid (direct audio input, silhouette inductors, and teleloops). Probably the most convenient coupling arrangement would be these last two possibilities which, however, require that the hearing aid contain a telecoil (see Chapters 2 and 9).

8. Recently, a universal receiver has been introduced (see figure 10). By inserting different modules, the receiver can provide access not only to wide- and narrow-band FM signals, but also to inductive loop and infrared systems. An audio input module provides direct access from such devices as a television set or a stereo system. It

Figure 10. Universal receiver, the "Chorus." (Figure used with permission from Audiological Engineering Corp.)

can also be used as a traditional FM radio by inserting an additional module. A built-in microphone permits the unit to be used as a personal listening device. Although the concept of a universal receiver is appealing, as of this writing no information is available regarding its actual performance in real life.

COMPARISON OF WIDE-AREA LISTENING SYSTEMS

Listening Comparisons

Nabelek, Donahue, and Letowski (1986) compared conventional PA systems to FM, infrared (IR) and induction loop (IL) systems in a medium-size classroom. There were four groups of subjects: young adults with normal hearing; two differing age groups who had mild-to-moderate hearing losses, but were not hearing aid users; and a fourth group of moderately hearing-impaired hearing aid users. Speech perception scores were obtained under two signal-to-noise (S/N) conditions (plus 8 dB and plus 20 dB).

The results show that for both S/N conditions the speech perception scores obtained with the FM, IR, and IL systems were generally superior to those obtained with the PA system. As the listening conditions worsened, the differences in scores between the PA system and the other three tended to increase. Minor differences in speech perception scores were found between the FM, IR, and the IL systems, but these were not considered to be clinically significant. The authors conclude that all three listening systems were equally effective in overcoming the acoustic degradations occurring in classrooms.

In a companion study, Nabelek and Donahue (1986) compared speech perception scores obtained with a PA system to those obtained with an FM and IR system in a large auditorium. They reasoned that the results obtained in a classroom may not have been applicable to larger listening areas. Five groups of subjects were tested. In addition to the four types of groups tested previously, Nabelek and Donahue (1986) added a group of non-native listeners with nonimpaired hearing. Two listening positions were chosen, representing a good (tenth row, center) seating position and a poor one (last row at the side under a balcony overhang).

The results basically corroborate those obtained in the earlier study. The speech scores achieved with either the FM or the IR system, for all groups in both positions, were superior to those attained with the PA system. Differences between the FM and IR systems were minor and not clinically significant. Either one could provide a listening situation significantly superior to that achieved with the PA system.

The results of the study permit a number of other interesting conclusions: first, in a large auditorium, unlike a smaller classroom, the speech perception scores of nonhearing-impaired people may be somewhat depressed while using a PA system. It should be recalled that this was also observed in the Bankoski and Ross (1984) study. Second, the experimenters were able to show that both an FM and an IR system could improve the speech perception scores of even non-hearing-impaired people in a large auditorium. Third, unexpectedly, the scores obtained with the PA system in the "better" seating position was worse than those attained in the "poorer" position for all groups participating in the study (also found in the 1984 Bankoski and Ross study). Therefore, we can conclude that we should not assume that what seems to be most appropriate seating in an auditorium upon visual inspection will be corroborated in reality. We *can* assume, however, that a listening system that bridges the distance from source to listener *will* improve speech perception, no matter where (within very broad limits) a person (nonhearing-impaired or hearing-impaired) sits in an auditorium.

Functional Comparisons

A question that has arisen frequently, and one that will be asked even more frequently in the future, is which of the three systems discussed in this section (IL, IR, or FM) is "best?" To answer this question insofar as FM and IR systems are concerned, one company that deals with both types prepared table II. Below I emphasize those points that seem most salient, and then compare the two systems to an IL system.

The major advantages of an IR system are that the same receiver can be used in any location that provides an IR transmission, any number of IR systems can be used in adjacent rooms, and they are impervious to structural and radio interference. Installing IR systems, however, is a bit more cumbersome, somewhat more expensive, and requires more expertise (to ensure coverage of an entire listening area) than installing a wide-area FM system. The "line-of-sight" reception for an IR system is an occasional annoyance, as is the necessity to keep the diode "bulb" uncovered.

FM systems, on the other hand, require little or no installation, and thus can be easily moved to different locations, the radio signal can "cover" even the largest auditoriums, and, used in conjunction with a personal transmitter, such systems can be employed in both large and smaller auditoriums, as well as outdoors. Because radio signals are not contained within walls, privacy will be compromised; if one does not want unauthorized "eavesdropping," an FM system would not be an appropriate choice. Reception may be affected by

Table II. FM or Infrared: Choosing a Hearing Assistance System

FM Systems	Infrared Systems
Advantages	**Advantages**
FM transmission easily covers any auditorium at a lower cost of coverage per square foot than infrared	Infrared transmission is confined inside opaque walls—you can use any number of systems in adjacent auditoriums without spillover
Multiple channels are available for multiple auditoriums	Infrared provides a secure transmission, if security is an issue
The channels can be changed to avoid external radio interference	Infrared is unaffected by external radio interference and cannot cause radio interference
The FM receiver "capture effect" can be used to minimize the number of channels needed for multiple auditoriums	The same receiver will work in any auditorium and pick up the correct program
Multichannel receivers are available to simplify receiver management	Any number of receivers can be used
Any number of receivers can be used	Infrared allows generally unrestricted seating
FM allows generally unrestricted seating	Infrared offers excellent sound quality
The receiver can be placed in a pocket or covered up	Hearing aid wearers can be accommodated
Persons or other obstructions don't block the transmission	
Hearing aid wearers can be accommodated	**Disadvantages**
FM is easier to install than infrared.	Smaller area of coverage than FM—higher cost per square foot of coverage
The transmitter can be concealed or covered	Infrared is usually more difficult to install than FM—transmitter placement is very important
FM is more portable than infrared	Infrared transmitter panels cannot be concealed or covered
FM offers excellent sound quality	The infrared receiver "eye" cannot be covered up
FM can be used in strong sunlight (outdoors)	Persons or other obstructions can block the infrared transmission
	Infrared is not as portable as FM
Disadvantages	Infrared cannot be used in strong sunlight (outdoors)
FM can be affected by external radio interference or cause radio interference	
There are a limited number of channels available	
The receiver must match the transmitter channel being used	
Receivers can pick up other auditoriums; with multichannel receivers, the user can select the wrong channel	
FM transmission is not secure if privacy is an issue	

Each system has unique characteristics and tradeoffs. For a given type of use, these characteristics will be a help, a hindrance, or they won't matter. This table details the key advantages and disadvantages of each system. In most cases, either type of system will work fine. Factors such as cost, radio interference, or area of coverage typically will make one system more suitable for your particular needs. Used with permission from Williams Sound Corporation.

competing radio signals or other types of electromagnetic disturbances. The availability of multiple channels can be both an advantage and disadvantage, depending upon the logistical difficulties in caring for and dispensing the appropriate receivers (presumably multifrequency).

Compared to the two previous systems, the major advantage of the IL system rests in its convenience; if a subject already uses a hearing aid that incorporates a telephone coil, then the "assistive receiver" is already in place. The signal is picked up by a personal hearing aid that, if the aid (and "T" coil) has been properly selected, imposes on the signal the desired electroacoustic characteristics. Those personal hearing aids with "M/T" switches (with both the microphone and the telecoil activated) provide access to *both* electromagnetic *and* acoustic signals.

The problem is that only a minority of hearing aid users (perhaps about 20%) possess hearing aids that include "T" coils. There seems to be a basic conflict between, on the one hand, the desire of many hearing-impaired people to wear the smallest, most invisible hearing aid possible and, on the other hand, the physical constraints imposed on the size of a hearing aid when a "T" coil is incorporated. Unless more hearing aids include "T" coils (preferably with the "M/T" option), the full potential of IL systems will not be realized.

IL systems, as described by Lederman and Hendricks (see Chapter 2, this volume), have essentially eliminated many of the problems occurring in earlier generations of IL systems. Spillover is minimal, equal field strength can be obtained, and positional orientation of the "T" produces only a minor effect. Although portable IL systems can be used, they will not produce the electromagnetic flux described by Lederman and Hendricks. The newer IL system requires a permanent installation because the loop is contained in a mat placed on the floor. In large auditoriums, therefore, only a portion of the floor space would probably be looped. Strong electromagnetic energy in the vicinity will interfere with the reception of the desired signal. Once in place, the IL system can be activated at the same time as the PA system and become a permanent feature of the listening environment.

Judging from past experiences with hearing aids, and just about all other kinds of electronic equipment, we are not likely to see any of these systems adopted universally. Market forces more than logic or performance, will ensure that potential purchasers and users will be required to choose from among these three systems. Fortunately, as reviewed earlier, any one of them is capable of improving speech perception beyond that possible with a PA system (and certainly superior to what can be obtained without any assistive listening device at all).

ACQUIRING A PERSONAL RECEIVER

There is one problem common to all large-area listening systems (with the exception of hearing aids with "T" coils), and that pertains to the care and dissemination of the receivers. Managers of large-area listening locations may be receptive to installing such a system, particularly one that is activated concurrently with an existing PA system; however, once they learn what is involved, their amenability may soon dissipate. When we consider the problems they will face in ensuring that one of their employees must check the receivers in and out, and ensure that the receivers are charged and in operating condition, then we can hardly fault their reluctance. The answer to this dilemma is for each hard of hearing person to possess a personal receiver for use in large-area listening systems.

This condition is met by hard of hearing people who use personal hearing aids with a "T" coil in places that provide an IL system. Except for in-the-canal (ITC) hearing aids, effective "T" coils can be incorporated in all other personal hearing aids. The advantages of using a personal hearing aid as a large-area listening device is that the individual who owns the hearing aid knows how to take care of it. The hearing aid need not be "checked" out and in, and the owner of the aid is responsible for ensuring that it is functioning properly. Even with the widespread adoption of "T" coils in hearing aids however, and the most optimistic forecast for the expansion of looped facilities, there will still be many large-space listening situations in which IL systems are either inappropriate or not installed.

If all indoor large-area listening locations provided an IR system, this would meet the requirement of a universal receiver. IR receivers are relatively inexpensive and can be used in any location that provides an IR transmitted signal. In some large cities (New York being one) IR systems are the ones used in most, if not all, the theatres that provide assistive listening systems. Under these circumstances, it is advantageous for hard of hearing people to purchase their own IR receiver. They then will be the agents responsible for ensuring that the system is operating correctly. However, many other large-area listening locations use the kind of FM systems discussed in this chapter. So what is the average hearing-impaired person to do? It is not economically feasible, or convenient, for the average hearing-impaired person to own different types of receivers for *each* type of listening system.

A common experience of many people with hearing losses is the malfunction or nonavailability of appropriate receivers in locations that presumably provide assistive listening services. The personnel responsible for charging the units either "forgot," or were never told of their responsibility; the units got misplaced somehow between performances; or somebody inadvertently put them under the heating system, or hosed them down while cleaning the facility. Taking care of the special receivers will never be a high priority function in locations that provide (with reluctance, under duress, or casually) a large-area assistive listening device.

The best solution at present is for each hearing-impaired person to own a universal receiver, of the kind described earlier (number 8 in the list of FM receiver features, above). This will accomplish two objectives: first, the unit can provide access to the complete range of current large-area listening systems. Second, it places the major responsibility for ensuring a successful listening experience on the person who has the most at stake—the one with the hearing loss. Because it is unlikely that any one kind of large-area listening system will capture the entire market, the universal receiver seems to be the best personal option for most hearing-impaired people.

REFERENCES

Bankoski, S. M., and Ross, M. 1984. FM systems' effect on speech discrimination in an auditorium. *Hearing Instruments* 35(7):8–12, 49.
Boothroyd, A. 1992. The FM wireless link: An invisible microphone cable. In *FM Auditory Training Systems: Characteristics, Selection, & Use*, ed. M. Ross. Timonium, MD: York Press.
Nabelek, A. K., and Donahue, A. M. 1986. Comparison of amplification systems in an auditorium. *Journal of the Acoustic Society of America* 79(6): 2078–2082.
Nabelek, A. K., Donahue, A. M., and Letowski, T. R. 1986. Comparison of amplification systems in a classroom. *Journal of Rehabilitative Research and Development* 23(1):41–52.
Ross, M. 1992. Room acoustics and speech perception. In *FM Auditory Training Systems: Characteristics, Selection, and Use*, ed. M. Ross. Timonium, MD: York Press.
Sergeant, L., Atkinson, J. E., and Lacroix, P. G. 1979. The NSMRL Tri-Word Test of Intelligibility. *Journal of the Acoustic Society of America* 65:218–22.

APPENDIX: PROVIDERS OF FM LISTENING SYSTEMS

Audiological Engineering Corp.
35 Medford Street
Somerville, MA 02143
(617) 623-5562

AVR Sonovation Inc.
1450 Park Court
Chanhassen, MN 55317
(612) 470-6633

Chaparral Communications
2450 North First Street
San Jose, CA 95131
(408) 435-1530
Comtek Inc.
357 W. 2700 S.
Salt Lake City, UT 84115
(801) 466-3463
Phonic Ear Inc.
3880 Cypress Drive
Petaluma, CA 94954-7600
(707) 769-1110

Sennheiser Electronic Corp.
6 Vista Drive
Old Lyme, CT 06371
(203) 434-9190
Telex Communications Inc.
9600 Aldrich Avenue South
Minneapolis, MN 55420
(612) 884-4051
Williams Sound Corp.
10399 W. 70th St.
Eden Prairie, MN 55344-3459
(612) 943-2252

Personal Listening Systems

Chapter • 5

FM Personal Listening Systems

EllaVee Yuzon

Assistive listening devices (ALDs) for hearing-impaired individuals have gained popularity over the past several years. Passage of the Americans with Disabilities Act has increased consumer awareness, and hearing professionals are providing "auxiliary aids and services" to their hearing-impaired patients. More and more hearing-impaired adults are benefitting from FM (radio) technology. This chapter focuses on this unique assistive device and its many uses.

DESCRIPTION OF AN FM SYSTEM

FM is a term applied to a radio wave that is frequency modulated (FM) by an audio signal emanating from a microphone. The radio FM signal is transmitted to a receiver designed to detect the desired carrier frequency. This signal is demodulated, amplified, and delivered to a listener (figure 1). In 1982, the Federal Communication Commission (FCC) authorized the use of 72–76 MHz as the designated bands to be used by persons with a hearing loss (tables 1 and 2).

Typically, FM technology is useful in the classroom as an educational tool for hearing-impaired students (Ross 1992). However, a "handicapping environment" is not limited to academic settings. Wherever acoustic information is affected directly by adverse listening situations or

Table I. Personal FM/Auditory Training (Channel Frequency Designations 72–76 MHz band)

Narrow-band Channels

Center Freq.	Biocoustics	Telex	HC-Phonic Ear Case/Dot	Comtek Earmark	Oticon	Zenith ZA-660
72.025	1	1	Red/Gray	1		1
72.075	2	2	Brown/Gray	2		2
72.125	3	3	Red/Brown	3		3
72.175	4	4	Brown/Red	4		4
72.225	5	5	Orange/Gray	5	Exact	5
72.275	6	6	Brown Orange	6	frequency	6
72.325	7	7	Orange/Brown	7	marked	7
72.375	8	8	Brown/Yellow	8		8
72.425	9	9	Yellow/Gray	9	on	9
72.475	10	10	Brown/Green	10	transmitter	10
72.525	11	11	Yellow/Pink	11		11
72.575	12	12	Brown/Blue	12	or	12
72.625	13	13	Yellow/White	13	receiver	M
72.675	14	14	Brown/Pink	14		14
72.725	15	15	Green/Gray	15		15
72.775	16	16	Brown/White	16		16
72.825	17	17	Green/Brown	17		17
72.875	18	18	Black/Gray	18		18
72.925	19	19	Green/Red	19		19
72.975	20	20	Black/Brown	20		20
TV-Audio band						
75.425	21	21	Black/Orange	21		21
75.475	22	22	Green/Yellow	22		22
75.525	23	23	Black/Yellow	23		23
75.575	24	24	Green/Blue	24		24
75.625	25	25	Black/Green	25		25
75.675	26	26	Green/Pink	26	Exact	26
75.725	27	27	Black/Blue	27	frequency	27
75.775	28	28	Green/Black	28	marked	28
75.825	29	29	Black/Pink	29		29
75.875	30	30	Pink/Gray	30	on	30
75.925	31	31	Black/White	31	transmitter	31
76.975	32	32	Pink/Yellow	32		32
74.625			White/Brown	33	or	
74.675			White/Red	34	receiver	
74.725			White/Orange	35		
74.775			White/Yellow	36		
75.225			White/Green	37		
75.275			White/Blue	38		
75.325			White/Pink	39		
75.375			White/Black	40		

by a hearing loss, the FM advantage should be sought. FM technology affords everyone the opportunity to participate in and enjoy the benefits of improved communication in every listening environment.

Table II.   Auditory Training Equipment (Channel [frequency] designations 72–76 MHz band)

	Wide-Band Channels		
Center Freq.	Phonic Ear/ Comtek	EFI	Zenith ZA-330
72.10	A (Red)	Yellow	A
72.30	B (Brown)	Orange	B
72.50	C (White)	White	C
72.70	D (Violet)	Blue	D
72.90	E (Yellow)	Green	E
75.50	F (Green)	Red	F
75.70	G (Black)	Silver	G
75.90	H (Blue)	Violet	H
74.700	I (Gray)		
75.300	J (Pink)		

In response to the needs for communication access of hearing-impaired people, FM technology can be found in the home, the workplace, and public and private facilities (figures 2 and 3). It has afforded

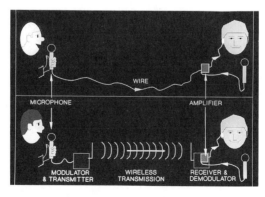

Figure 1.   The desired audio signal is imposed on a radio wave by a process of modulation. The signal is received and demodulated (the removing of the signal from the carrier wave). The signal is amplified and interpreted as sound by the listener through headphones, speakers, or other appropriate transducers. (Courtesy of Boothroyd.)

Figure 2.   Use of a Comtek assistive-FM system in the workplace. (Courtesy of Comtek.)

many hearing-impaired individuals the opportunity to be vital, contributing participants in their everyday communication.

FM RATIONALE

Typically, hearing aids alone cannot accommodate the total communication needs of hearing-impaired persons. A hearing aid is a good solution for sounds that are close; it does the job for which it is designed, but there are many shortcomings. Because hearing aids are designed to deal with problems of threshold elevation, their performance may be limited by acoustic feedback, variations of speech input levels, and the low signal-to-noise ratio for remote speech. Even state-of-the-art hearing aids cannot compensate for distortions in sound that occur as they travel from the source in a noisy and/or reverberant environment. The perception of a speech signal is further compounded by an individual's hearing loss, thus intensifying his or her difficulties in comprehending speech. Typically, to make use of their auditory

Figure 3.   Recreational use of an assistive FM-system. (Courtesy of Comtek.)

skills, hearing-impaired individuals must somehow overcome their poor listening environments. Usually, a listener will try to compensate by using visual clues such as lip reading, but these are of limited value in many listening situations.

For people with normal hearing, a signal-to-noise ratio (S/N) of +6 dB typically allows for the reception of intelligible speech. Persons with a hearing loss usually need an S/N of about +20 dB (Finitzo-Hieber and Tillman 1978). The preferred signal must be "sufficiently" louder than the background noise. It must bridge the distance between talker and listener. FM technology is the solution.

FM TECHNOLOGY

The use of an FM system designed for hearing-impaired individuals has several advantages over conventional amplification. First, a microphone can be placed close to the originating signal. The quality of the signal is determined by the nature of the source and the placement of the microphone near that source. Placing a microphone six to eight inches from a speaker's mouth or from the desired signal gives a listener optimum access to critical speech information (figure 4). Using different types of microphones and the placement of those microphones can enhance a selected signal from +15 dB to +25 dB signal-to-noise ratio.

FM systems can be coupled directly to the "audio out" plug of any audio source, such as a TV or VCR, with remote microphones or direct audio connection. An FM system allows the distance between a transmitting signal and a receiver to be bridged without any loss of integrity even in the presence of noise. The use of an FM system

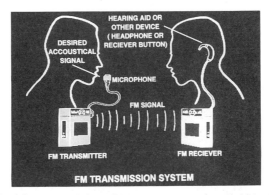

Figure 4.   Use of a remote microphone not greater than 6–10 inches from the desired audio source. (Courtesy of C. Compton, Gallaudet University.)

makes it possible for hearing-impaired adults to benefit from a higher input level and an improved signal-to-noise ratio, ensuring optimum comprehension. An FM system is, thus, an essential tool for the acquisition of information by adult users in many everyday communication situations.

TYPES OF FM SYSTEMS

Manufacturers have developed a variety of FM systems to address various fitting needs, budgets, and applications. Each system has its own merit and purpose in addressing the communication needs of the hearing-impaired population.

Self-Contained FM Systems

Personal FM systems are used with or without hearing aids. Some FM manufacturers have designed self-contained FM systems to be fitted as a hearing-impaired person's main amplification. Binaural or monaural systems are available with tone, gain, and output controls similar to those found in a conventional hearing aid (figure 5). The packaging includes transmitter, receiver, and one of four coupling options for use with the self-contained receiver (figure 6). Typically, this system is fitted on children, predominantly in schools where children do not have hearing aids or wear them inconsistently. The advantage for the school is the ability to give each hearing-impaired child the added benefit of FM.

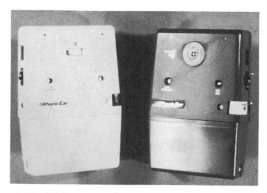

Figure 5.   Phonic Ear FM system. Contains both FM and HA circuits. (Courtesy of Phonic Ear.)

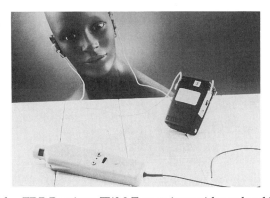

Figure 6.   Telex TDR Receiver, TW-3 Transmitter with ear-level BTE transducers. (Courtesy of Telex Communication, Inc.)

Self-Contained BTE/FM

The recently introduced behind-the-ear (BTE)/FM receiver (figure 7) is the same size as many conventional BTE aids. It functions as an individual's main amplification with the added advantage of FM technology as an integral part of the system. The options of hearing aid (HA) only, HA with FM, and FM only easily accommodate the listener in a variety of listening environments. The prescribed settings can be identical in both the HA and FM positions. In addition the FM transmitter is equipped with a high-frequency tone control. Also included is a gain control on the FM transmitter for adjusting the sensitivity of the transmitter microphone. This gives a listener the flexibility of adjusting from one-to-one selective listening to possible group access. Binaural and monaural systems are available with a matching, single-channel FM transmitter, charger, and carrying case.

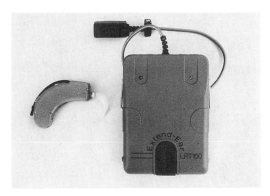

Figure 7.   Extend-Ear BTE/FM System. (Courtesy of AVR/Sonovation.)

Uses for this type of self-contained BTE/FM system expand beyond schools to include occupational and social situations. It is a possible choice for adults who are in the market for both hearing aids and an FM system. However, the listener's demands, amount of hearing loss, and the need for compatibility with other FM systems need to be considered.

FM SYSTEMS WITH PERSONAL HEARING AIDS

Many adult listeners may like their current hearing equipment, but there are times when they also need the benefits of an FM system. This puts them in the market for an FM system designed to enhance their auditory functioning in certain adverse listening situations. The assistive-personal FM system operates in conjunction with the listener's personal hearing aid coupled by an appropriate cord or transducer (figure 8). Typically this FM system is packaged with a transmitter and microphone, a personal receiver with the appropriate transducer, rechargeable batteries, charger, and carrying case.

Many assistive FM systems provide an optimum signal, surpassing the performance characteristics of a hearing aid. A hearing aid and its prescribed settings, thus, determine the amplified signal arriving at a listener's ears. Careful attention must be paid to the many cord and transducer variables involved when coupling an FM receiver to a personal hearing aid (Thibodeau and Saucedo 1991). Acoustic and behavioral information should be evaluated to ensure that the optimum signal is received by a listener. By doing so, one can be moderately assured that, as long as the proper hearing aid, earmold fitting, and coupling option are selected, the same "prescribed" amplified speech

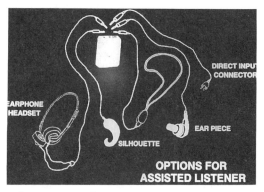

Figure 8.   Transducer options for use with assistive-FM systems.

signal will be delivered to benefit a hearing-impaired individual (Seewald and Moodie 1992).

CHOICES AND OPTIONS

Basic FM Systems

Basic FM systems are designed simply for the transmission and reception of an FM signal. The package usually includes a transmitter, microphone, receiver, transducer, and charger. These systems vary in cost based upon their specifications, workmanship, and quality. They are typically designed for listening to a single source, such as a speaker in a church, a teacher, spouse, or lecturer (figure 9).

It is also important in many situations for the listener to monitor his or her environment. A basic assistive FM system should be linked with a hearing aid that includes an "MT" switch or one marked "+." This gives a listener the advantage of being able to switch the hearing aid from FM only ("T" position) to respond to both hearing aid and FM signals. If this option is not available, then an assistive FM system with a receiver with an environmental "mic" or "aux" input should be considered. Comprehension of speech in both quiet and noise should

Figure 9.   Communication access to a single speaker (i.e., tour guide, lecturer, teacher). (Courtesy of Sennheiser Electronic Corporation.)

be evaluated. Often, when noise and reverberation exist the benefit of FM is defeated or lessened with activation of an HA microphone.

Assistive FM Receiver with Environmental "Mic" or "Aux" Input

The receiver in this assistive FM package may include an internal or external environmental microphone. Typically, the microphone can be activated or deactivated by either a mute switch or by plugging in or removing the microphone. This gives listeners the ability to adapt and control their listening environment in a variety of situations. As an added benefit, some of these systems have trimmer pots or controls to adjust the gain of the environmental microphone in reference to the transmitter microphone (figure 10). This is a critical option, especially when signals emanating from the two microphones become competing signals, thus limiting the benefit of FM usage. The noisier the environment, the more likely it is that such competition will occur. By setting the gain of the environmental microphone, a listener can adjust the relationship between the two signals arriving at the ear. Listening environment and desired primary source should be evaluated to determine the relationship between the transmitter and environmental or hearing aid microphone outputs.

New Assistive FM Technology

In response to the challenge of using two microphones simultaneously, some FM manufacturers are incorporating a newly developed circuit that prioritizes the two competing signals. This feature automatically gives preference to a voice signal received from a transmitter.

Figure 10.   Independent gain controls to adjust the transmitted FM signal in relation to the environmental sounds. (Courtesy of Sennheiser Electronic Corporation.)

The environmental microphone's signal is suppressed by as much as 15 dB when an FM signal is transmitted. Because of the nature of the circuit, the automatic priority is usually not triggered by ambient or environmental noise. A sensitivity level is preset in the receiver to determine the intensity of voice versus noise. When no signal is being transmitted by the FM transmitter, the gain of the environmental microphone automatically readjusts to its previous level. This enables an individual to have access to environmental conversation from his or her surroundings. However, because the receiver always gives priority to the FM transmitter signal, and the environmental microphones are only activated in the absence of an FM acoustic signal, passive monitoring and awareness of sounds in a person's surroundings are reduced. This may be disturbing to those with a special need for monitoring their environment.

Extended Uses of Receiver with Environmental Mic/Aux Input

Another benefit of FM receivers with environmental microphones and/or "aux" input is the capacity for the receiver to serve as an independent amplifier. The receiver can be used independently from the transmitter to provide access to a recorded signal, a computer language program, a telephone, or to monitor the user's own speech (figures 11 and 12). Distance may not be a factor in a listening situation, thus eliminating the need for a transmitter. When a receiver is operating as an amplifier, the frequency selector should be removed. This prevents the listener from hearing "white noise" or receiving interference from another radio broadcast. If the frequency selector/crystal cannot be removed, as with internal frequency controlled receivers,

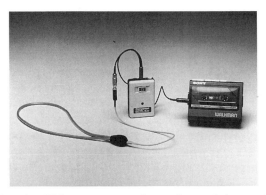

Figure 11.   Use of an FM system to monitor and record the speaker simultaneously. (Courtesy of Comtek.)

Figure 12.   Use of an FM receiver as amplifier only with the telephone. Provides additional amplification, enhanced frequency response, and binaural listening.

the transmitter's microphone should be muted or disconnected. The transmitter continues to emit radio signals to an assigned receiver without the audio to interfere with the preferred signal.

Other options and packaging should also be considered in selecting a personal assistive FM system. There is the choice of channels, wide band versus narrow band (table III). FM systems can be purchased as fixed-single channel, tunable multichannel, multichannel digital, multichannel crystal controlled, and so forth. The need to be compatible with other systems, the quality of the signal influenced by the type of channeling, as well as the ease in selecting or tuning to channels should be considered.

MICROPHONES

There are just as many choices of microphones: lapel, internal, lavaliere, plug-in, boom, tabletop, desk stand, head worn, and so forth (figures 13 and 14) with varying kinds of operational characteristics (omnidirectional, unidirectional, pressure-zone, noise-canceling, etc.). The use of one microphone may be appropriate for one application but require a different microphone for another application. Each person's different listening environments should be considered in the selection process. The greater the hearing loss, the greater the need for a higher signal-to-noise ratio which usually means placing a microphone near the speaker's mouth. Headworn, boom, and/or noise-canceling microphones are designed to produce a maximum signal-to-noise ratio. However, the microphone delivering the best S/N is not always cosmetically preferred (figure 15).

Table III.   Advantages and Disadvantages of Narrow-Band versus Wide-Band and Tuning Options

	Advantages	Disadvantages
Narrow-band channels 1–40	Can accommodate more systems within the same parameters or building. 40 channels available Less chance of interference	Narrower dynamic range Limited frequency response May clip the performance of some high frequency emphasized hearing aids
Wide-band channels A–J	Wider dynamic range frequency response of most conventional hearing aids and technologies.	Greater deviation that may result in greater chance of interference Limited to 10 channels
Tuning Options:		
Single-channel fixed-crystal controlled	Ensure capture and lock-on to matching channel	Internal crystal limits usage to assigned channel only
Single-channel with external frequency selector	Ensure capture and lock-on to assigned/matching channel Flexibility of plugging in and matching the personal receiver to all 40 narrow-band channels or all 10 wide-band channels depending on the band width	Individuals with poor dexterity may have difficulty removing and replacing frequency selectors The inconvenience of having to carry additional selectors.
Multi-channel tunable	Usually Wide Band with wide frequency response Does not require carrying frequency selectors Usually inexpensive	Deviation between 50–75KHz Greater chance of interference Voice is difficult for many hearing-impaired individuals to tune to, unless used with base station transmitters with tone generator for easy tuning Has a tendency to drift Does not click or lock onto channel
Multi-channel crystal controlled (selectable)	Available in wide-or narrow-band channels Flexibiity in matching receiver to many other systems for educational or ADA applications	Individuals with poor vision and poor dexterity may have trouble selecting appropriate channel
Multi-channel digital	Lock/click onto assigned channels Quality reception Low interference	Difficult for those with low vision and/or poor dexterity

Figure 13.   Omnidirectional lapel microphone and deskstand microphone samples. (Courtesy of Phonic Ear.)

TRANSDUCER OPTIONS

A personal FM receiver can be worn with a variety of devices (i.e., in-the-ear [ITE], behind-the-ear [BTE], cochlear implants, vibro-tactile devices, frequency transposition amplification, etc.). A personal receiver can transfer a received signal to a listener's ear through one of several options.

Walkman

Walkman-type headphones and earbuds are typically used by those people without hearing aids. However, Walkman-type and similar headphones can be used in conjunction with canal or ITE hearing aids.

Figure 14.   Tabletop pressure zone microphone and headworn noise-canceling microphone. (Courtesy of Comtek.)

Figure 15. Headworn noise-canceling microphone. (Courtesy of Comtek.)

The microphone of an aid receives the signal that the hearing aid shapes and amplifies as prescribed (figure 16).

Button Receivers/Earphone Assemblies

Button receiver/earphone assemblies can be used with either a universal fit eartip or coupled to a custom snap-on earmold. Custom earmolds create less feedback because of the increased distance from the microphone and the secure fit. A variety of button receivers, differing in frequency response, efficiency, and output are available. The majority are very efficient and are designed for high-level output. In these cases, caution should be exercised to ensure that an individual does not receive over-amplified sound. Attenuation to limit the output can be accomplished by trimmers on some FM receivers, by changing to a

Figure 16. Use of an FM receiver with Walkman-style headphone. (Courtesy of Comtek.)

less efficient button receiver or transducer, or by using a cord modified with resistors (figure 17).

It is important to note that all headphones, earbuds, button
receivers, and earphone assemblies are not equal. Manufacturers can
help recommend and outline required transducer specifications compatible with their systems. Inappropriate transducers and mismatched
impedances may result in audible distortion.

Neckloops and Silhouettes

Neckloops and silhouettes convert an audio signal from an FM
receivers' output into a magnetic signal (figure 18). They create an
electromagnetic field that the hearing aid, in the "T" (telecoil) position,
detects and converts back to electrical energy. This signal is then
amplified and shaped by the hearing aid. The choice of neckloop or
silhouette should be determined by the performance of the hearing aid
and the efficiency of its telecoil (table IV and V). Also, the frequency

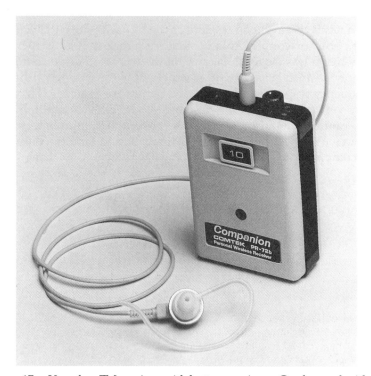

Figure 17.   Use of an FM receiver with button receivers. Can be used with
universal eartips or custom snap-on ear molds. (Courtesy of Comtek.)

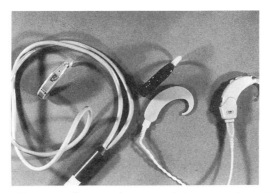

Figure 18.   Neckloop or silhouette for use with hearing aids with telecoils.

characteristic of the hearing aid in the "T" position may vary from the prescribed response in the "M" position (see Chapter 2). The appropriateness of neckloops and silhouettes over other coupling options and the performance and preferences of the individual should be considered. It is important to note that when a hearing aid is placed in the "T" position, the hearing aid microphone is deactivated. In some listening situations that may be preferred, but when the individual needs to monitor the environment, it is critical to have a "MT" switch or "+" (telecoil plus hearing aid microphone).

Also to consider, many listening environments have levels of magnetic fields that might be detected by the telecoil. Computers, fluorescent lighting, microwaves, and other common home and office appliances may produce such magnetic interference.

Direct-audio-input

Direct-audio-input (DAI) is an electrical signal sent directly to a hearing aid via a cord coupled to the aid's shoe or boot adapter (figure 19). Verification of the performance of a hearing aid with the direct input cord is recommended. Direct input should not fundamentally alter the electroacoustic performance of a hearing aid.

Table IV.   How to Use a Telecoil Successfully With An Assistive -FM System Using a Neckloop or Silhouette (Courtesy of C. Compton, Galludet University)

1.   Telecoil with sufficient sensitivity
2.   Correct orientation of telecoil within hearing aid chassis
3.   Adequate magnetic field emanating from the neckloop or silhouette
4.   "M/T" toggle switch, if possible
5.   Client instruction on proper use of telecoil
6.   Absence of electromagnetic interference from computers, motors, etc.

Figure 19. FM directly coupled to user's hearing aid with the appropriate cord and audio-shoe (boot).

Specialty cords and transducers have been designed to use FM systems with implants, vibrotactile devices, bone conductors, frequency transposers, etc. (figure 20). As with hearing aids, these devices are designed to receive sounds via built-in microphones. Unfortunately, speech perception through these special devices (as in hearing aids) is affected by background noise, reverberation, and distance. Fortunately, these devices can be interfaced with FM technology to improve the quality of signal. An FM signal is coupled directly to the device by an appropriate connector, especially designed by the device manufacturers. Cochlear implant processors and other devices are not designed to accept high voltage. The use of a generic cord in linking an FM system to these systems may damage the equipment. When linking devices for the first time, it is wise to put the volume control on the FM at the minimum setting and increase as needed.

The remaining consideration is the need or limitation of an FM system's range, typically, up to 300 feet. However, some systems'

Table V. How to Improve Telecoil Strength (Courtesy of C. Compton, Galludet University)

- Order telecoil equipped with preamplifier circuit (Not all companies will do this)
- If faceplate is too small, wire two telecoils in tandem (Add 6dB)
- Add more coil of wire around telecoil's ferriferous core
- Increase length of the telecoil's ferriferous core

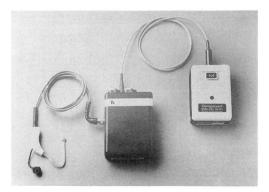

Figure 20.   Use of FM system interfaced with the processor of a cochlear implant. FM can also be coupled to vibrotactile, frequency transposers, and bone conductors.

ranges may be less, limiting communication access in certain large areas. Finally, the cost for an FM system ranges anywhere from $300 to $3,000. The price usually reflects the system's packaging, utility, quality and options.

THE PROFESSIONAL'S ROLE

Personal FM systems represent a major breakthrough for people with hearing losses. The hearing professional's recommendation of an FM system must be with the user's specific needs in mind, and the client should play an active role in selecting a system that may be most appropriate to meet his or her communication needs. To make an informed decision, users must have an understanding of the advantages and disadvantages of each system. They should be made aware of the trade-offs associated with choosing cosmetics over function. The best acoustic result may dictate the use of direct-audio-input over magnetic induction. With the help of a hearing professional, users need to consider and prioritize the factors listed in table VI.

   After hearing-impaired individuals make their choice, it is critical for them to learn, with the help of their hearing professional, how to use their FM system in all listening applications. FM users should become knowledgeable enough about their equipment to use it whether for interpersonal communication, reception of broadcast media, telephone communication, speech recognition in large meetings, receptive conversation in groups (family, friends, colleagues, travel companions, medical personnel, and so forth) (figure 21 and 22). Feeling comfortable with their equipment encourages use of the system, wherever it is needed.

Table VI. Factors To Consider in Selecting an FM System

• Acoustical performance of the hearing aid	• Effectiveness
• Coupling options	• Portability
• User lifestyle	• Cosmetic
• Environments where the FM system is to be used	• Compatibilty
	• Dependability
	• Operability
• Audio sources to which the FM needs access	• Durability
	• Cost

Whether an adult chooses an FM system as an addition to standard amplification or as primary amplification, there are many successful and valuable uses for an FM system. The following cases reflect various uses for a personal FM system in everyday listening situations.

FM APPLICATION

Case #1

Attending church and hearing church services was important for a grandmother of eight. But it was becoming more and more difficult for her to comprehend voices in what she perceived as a very noisy environment. She chose an FM system to interface with her hearing aids. Now she simply hands the transmitter and lapel microphone to the pastor who clips the microphone to his lapel prior to services. By using an FM system, she is again able to feel like a participant in her neighborhood church and to enjoy the service. In fact, she sat in the

Figure 21.   Use of an FM with a tabletop microphone for communication access of a group.

Figure 22.   Communication access in a board meeting.

last row during an overcrowded Easter service and heard every word. The pastor later learned that there were many other members who were having difficulty hearing in church. He is now looking for funds for a large-area FM system.

Typically, the adult FM user attending a meeting in an auditorium that does not have an ALD simply hands the speaker an FM transmitter and microphone prior to the meeting. When there are multiple speakers, and to alleviate the handling and passing of the FM transmitter, the user can simply place his or her microphone near, or attach it with velcro to, the podium microphone, where it is near the main speaking position of the presenters. A third option is to connect it temporarily to the audio-out plug of the existing lectern or public address system, especially if more than one microphone is used during the service or proceedings. Some personal FM systems include an audio cord for such applications. Often, that is the preferred option; however, it does require some time, the appropriate cord, and assertiveness in hooking it up to the sound system.

Case #2

A hearing-impaired woman enjoys going to the movies with her family. The theater does not have an assistive listening device for her to use. She informs the theater management of her needs ahead of time. The management agrees to hook-up her FM to the "audio/phono" out of the sound system of the selected movie. She arrives early to ensure seating and to hand her transmitter and audio patch cord to the manager.

It is important to note that her system's antenna for long range is an integral part of her lapel microphone. The use of her personal transmitter with the audio patch cord had limited range but was sufficient in reaching her seat in the theater. If the transmitter's location had been at a greater distance, her success might have been limited. The range of an FM system with an external antenna is not affected when coupled to an audio source. However, a base station FM transmitter would be used for a more permanent installation (see Chapter 4).

Case #3

A hearing-impaired individual is taking a refresher course on the new tax laws. A lot of information is being given. She is having a difficult time with all the note taking that is required. To relieve the pressure of straining to hear and take notes simultaneously, she equips her receiver with a special "Y" cord that allows her to listen and to tape the transmitted signal simultaneously. The black connector plugs into the "mic" input of the cassette recorder, with the red end connector plugging into the receiver's output. The neckloop (or other listening piece) plugs into the end of the silver female connector. Later, she is able to play back and listen to a quality recording, using her neckloop directly in the "headphone-out" on the recorder, or using the specialty cord with her transmitter coupled to the audio-out of the recorder. The red plug inserts into the output of the audio source of the cassette and the black plug inserts into the transmitter for transmission of the recorded information (figure 23).

Case #4

A hearing-impaired employee occasionally is required to work on a computer that uses an audio program (voice simulated). On the side of the computer is a box with an audio-out jack that allows the hook-up of headphones. To get the best reception from the computer's audio program, an audio cable was designed (matching impedance levels) to take the output from the "aux" output (headphone-output) into the environmental mic/audio input on the employee's personal receiver. The environmental/audio input gain was adjusted to optimize the level of the incoming signal for distortion-free sound. The listener adjusted the personal receiver's volume to a comfortable loudness level. Because computers emit a magnetic field, the hearing-impaired employee was fitted for direct-audio-input from the personal receiver to his hearing aids. An FM transmitter, with mute switch, was worn by the department director. The transmitter allowed him to maintain his "communication access" to the employee throughout the workday. The director would activate or deactivate the transmitter microphone by flipping a short toggle mute switch on or off. If the transmitter was not in use and the microphone not muted, and the hearing-impaired employee wanted to deactivate unwanted conversation, he simply

To transmit a program from your transmitter and any other audio source . . .

Insert black plug into M-72 transmitter microphone jack.

Insert red plug into output of audio source jack.

Results: A direct connection to what's being played by your tape recorder, TV, etc. If you are accessing a VCR, an RCA jack is required. (An adapter can be purhased at any audio shop.)

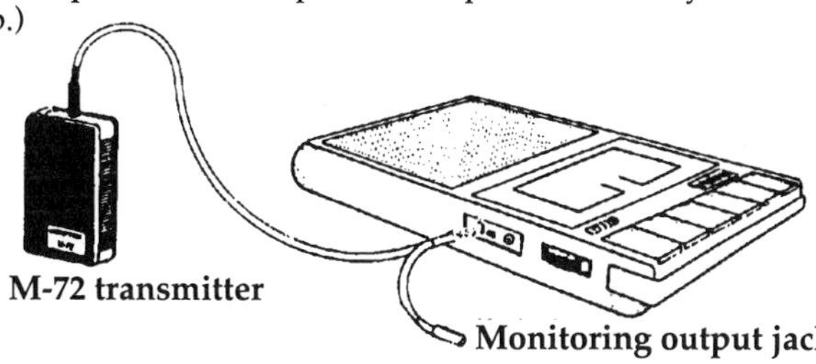

M-72 transmitter

Monitoring output jack

To Record and Monitor the Speaker Simultaneously

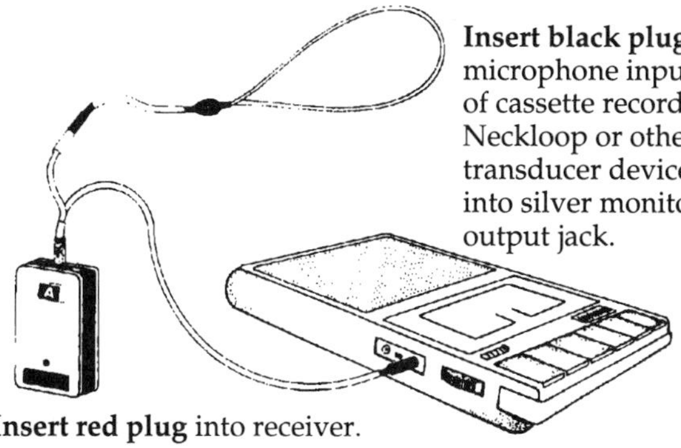

Insert black plug into microphone input jack of cassette recorder. Neckloop or other transducer devices plug into silver monitoring output jack.

Insert red plug into receiver.

Result: A direct line from the speaker to you and your recorder for playback at your convenience.

Figure 23. Expanding Your Connections. (Courtesy of Comtek.)

removed the frequency selector from the back of the personal receiver. The receiver then functioned as an amplifier only to the signal emanating from the computer.

Case #5

In a rehabilitation training program, the same concept was taken one step further. A hearing-impaired listener needed access to an audio program, the instructor, and her assigned computer aide. It was an extremely noisy environment because of the number of people and computers. Yet she needed to hear all three sources selectively. The instructor wore a transmitter with the mute switch in the same manner as the director did in Case #4. A "Y" cord was designed, allowing her to have access to the audio program of the computer, while the computer aide wore a noise canceling microphone and used headphones coupled to the secondary audio-out of the computer in order to track the audio of the program. The intent of the rehabilitation program was to have the hearing-impaired listener sharpen her listening skills from the audio information. The aid was to reiterate information about the task until the listener gained sufficient skill to complete the task independently. With the use of noise canceling microphones, a specialty cord, and a turn-taking protocol in speaking, the listener could meet her communication needs in the training program.

Case #6

An insurance agent's hearing loss had progressed to the point that communication with his clients in the office and on the telephone became extremely difficult. He felt that his clients' perception of his competence to do the job was being challenged by his hearing loss. Many times he used his secretary as an interpreter, but he felt communication was limited when others tried to represent him. He had become so frustrated that he considered taking an early retirement. When an FM system was demonstrated to him, he did not have much faith in the device, yet he felt it might be his last chance to become vital again in the workplace.

An assistive FM system with an environmental/audio input receiver was used. A line-level telephone adapter was coupled to his desk phone. The adapter did not change nor interrupt the signal feeding to the handset. A special cord took the audio signal from the adapter and was fed directly to the environmental/audio input of the receiver (figure 12). Again, the receiver acted as amplifier only, enabling the listener to hear telephone conversation binaurally through his hearing aids. (It is important to note that some telephones work better than others.)

A tabletop microphone was linked with his transmitter and positioned near the place where a client would sit. The type of microphone and its proximity enabled him to understand his clients. The use of an FM made all the difference; it helped him to regain confidence in his abilities.

Case #7

A podiatrist had difficulty hearing his patients' comments as he looked down at their feet. He felt awkward calling attention to this hearing impairment. Asking his clients to talk into an FM system was not acceptable to him. The answer that worked was to attach the microphone with velcro to the right side of a wing chair where his clients sat. He works toward the right side of the chair to ensure that a patient's head and mouth naturally turn toward the microphone. Because he is relying on auditory information alone when his head is down, it is critical to his livelihood to receive the best auditory signal possible while in this position.

Case #8

A judge was having increasing difficulty hearing in his courtroom. It was critical that he hear every comment, cross examination and objection. A public address system was in use but he felt he received little benefit from it. Voice-activated microphones were in place at the witness stand, on the judge's bench, and at a podium used by both attorneys during their examinations. The building was old with high ceilings. His courtrooms were noisy with a lot of reverberation. Also, the judge needed the flexibility of a system for use during private consultation within his chambers. A combination of a personal system adapted for use with a base station adapter was recommended. A base station adapter was permanently installed with the courtroom's public address system. The belt-pack transmitter simply slipped into a pocket on the back of the base station. When the transmitter is used with the base station, the battery is removed because it can then be AC powered. The transmitter is easily removed, coupled to either a lapel or tabletop microphone and powered by a 9-volt battery for interpersonal use. It thus acts as both a large-area system as well as a personal system. The judge uses a neckloop, worn under his robe, that magnetically drives the telecoil in his BTE hearing aids. His receiver has an environmental microphone that extends to the front of his bench. Using the environmental microphone in that matter is helpful when attorneys are asked to approach the bench. Depending on the listening situation, his FM system is flexible enough to meet his communication needs in the courtroom.

Case #9

An older couple saved for their retirement and periodically travelled across the states with another couple. Initially, the wife enjoyed the trips but was getting to feel less and less a part of the group. Because of her hearing loss, the noise level in the car and the distance from the couple in the back, she was rarely able to participate fully in the conversations. She preferred to sleep during the long drives rather than strain to hear or feel embarrassed when she responded incorrectly to topics or repeatedly asked, "What?" The woman's hearing spouse and friends valued her friendship, and it was important to them to be able to include her in their conversations. The solution was easy. An FM transmitter was placed in the back with the other couple. The transmitter was adapted with a "Y" cord and was attached with velcro to the back seat. The "Y" cord was designed with two female 1/8 connectors. Each connector was coupled with a unidirectional microphone that could be clipped to the lapels of the two speakers in the back. The transmitter was able to power efficiently up to two electret lapel microphones. To hear her husband, the woman's FM receiver's environmental mic/audio input was also adapted with a "Y" cord, coupled with two unidirectional microphones. One was clipped to her husband's clothing and the other one to her's so that she could monitor her own voice.

When the couples were on a guided tour, the transmitter unit with lapel microphone was worn by the tour guide, when on their own, by her husband. When dining, an omnidirectional microphone was placed strategically across from her, near the farthest speakers. Sometimes a flower arrangement made a convenient stand in which to clip the microphone. Her environmental microphone on her personal receiver picked up conversation near her.

Case #10

In this example, a hearing-impaired gentleman had many situations that necessitated the use of a personal FM. However, one difficult listening condition was created when watching television. Prior to his use of an FM system, the volume control of the TV was usually set at a level that was uncomfortable for other family members. Even with the TV volume turned up, the programs were not always more intelligible. His television was equipped with a VCR that enabled an audio cord to be connected to the "audio out" plug of the VCR. The cord was placed so that it was easily accessible. By using the FM transmitter directly coupled to the television/VCR he was able to hear the TV without disturbing others.

In some cases, certain television channels may be affected by interference from an FM transmitter. The problem is usually alleviated if the FM transmitter is placed a few feet away from the TV.

Some TV's are not equipped with a VCR or a cable box that can easily be accessed. The solution may be to use the "audio-out" plug directly from the TV. However, in some cases, the TV speaker will automatically deactivate and only the FM can receive and transmit any audio from the TV to the FM personal receiver. This is not a good solution for a family gathering. In such cases a lapel microphone is simply placed or attached with velcro on or near the television speaker. The volume control of the TV can be set at a comfortable, audible signal for everyone else. The television's volume setting is critical to the quality of sound received by the FM receiver. Volume that is too high may overdrive the transmitter microphone, which results in high distortion. Using the microphone to gain access to the speaker is not always best, basically because it is at the end of the audio line. But the quality is usually good enough to enhance one's perception and comprehension (figure 24).

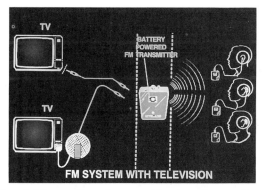

Figure 24.   Use of either a microphone or direct-audio coupling to access the audio of a television program.

Case #11

As she busied herself with daily chores, a woman caring for her ailing spouse could not always hear her husband's calls for assistance. This angered him and worried her. An FM transmitter with microphone was placed on the nightstand near his bed. When he would call to her, even when she was outside gardening, she was able to respond quickly to his needs.

Whether the communication needs are in the home, on the job, or of an education, recreation or traveling nature, the use of an FM system is invaluable. Often, it merely requires some creative thinking and a few trials to accommodate all of a person's communication needs. The Americans with Disabilities Act is expanding awareness and use of personal-FM technology.

IMPLICATIONS OF THE ADA

The Americans with Disabilities Act (ADA) is a landmark civil rights law that, for the first time, clearly entitles hearing disabled individuals to the same access to information, services, activities, and programs in their community, work, and educational environment as is afforded to every other American. Many public entities are seeing "communication access" as good business.

"Communication access" in public gathering places can be implemented effectively and affordably with permanent or portable FM assistive listening devices. As stated earlier, a person can gain access to the preferred sound source either by microphone or direct audio hook-up, thus sending the enhanced signal via FM to personal receivers in an audience. Multiple channels, such as in a multiplex theater, can operate simultaneously without interfering with each other. In some situations, multiple FM systems are required in a single meeting hall for visual narration, language translation, as well as access for the hearing impaired. Each system transmits on an independent channel, and the receiver's channels can be changed to accommodate easily day-to-day communication requirements.

Group FM systems can be permanently installed into an existing public address system in a large meeting place. Receivers may then be "checked out" by listeners with each unit tuned to the transmitting frequency. If there is an assembly of 50 or more people without the benefit of a public address system, the speaker simply uses a belt-pack transmitter with a lapel microphone for transmission. The ADA sets the number of receivers to be made available in large-area systems at four percent of the seating capacity. Multiple receiver units are stored typically in a twelve unit storage/carrying charger case or placed in charging modules that have the flexibility of add-on to charge up to twelve units.

By knowing the FM manufacturer and channels installed in public facilities, a person may be able to use personal receivers by simply selecting or tuning the FM system to the same channel. It is necessary to ensure that narrowband is matched to narrowband and wideband to wideband. However, the majority of FM systems installed in public areas are wideband channels. This may be a consideration in choosing a personal FM system and its channeling capability.

It is exciting to think how technological developments and the enactment of the Americans with Disabilities Act will improve communication and independence for persons with impaired hearing. Imagine on a typical day a hearing-impaired adult going to his or her place of employment and having access to information at a conference meeting with a personal FM system. He or she later takes business calls on a specially equipped telephone, enabling binaural listening. Later, he or she takes a break in the employee's lounge and is able to "mic" and participate in the table conversation with colleagues. Prior to going home, he or she stops at the county court to object to a speeding ticket. An FM assistive device is provided to ensure the individual access to the vital information to plead his or her case. Continuing home, he or she picks up the family and proceeds to the local movie, taking advantage of an FM system integrated with the sound system. Because the hearing-impaired person prefers his or her own FM receiver, he or she simply tunes it to the channel advertised on the sign near the entrance. After the movie, the family stops to discuss the movie and the day's events over a soda. Again the FM system is used to hear the family's voices.

Everyone should expect nothing less than such a "Typical Day"!

REFERENCES

Boothroyd, A. 1992. The FM wireless link: An invisible microphone cable. In *FM Auditory Training Systems: Characteristics, Selection, and Use*, ed. M. Ross. Timonium, MD: York Press, Inc.

Compton, C. 1990. Assistive devices, needs assessment and fitting protocol. Presented to the Wyoming Speech and Hearing Association.

Finitzo-Hieber, T., and Tillman, T. W. 1978. Room acoustics effects on monosyllabic word discrimination ability for normal and hearing-impaired children. *Journal of Speech and Hearing Research* 21:440–58.

Ross, M. 1992. Room acoustics and speech perception. In *FM Auditory Training Systems: Characteristics, Selection, and Use*, ed. M. Ross. Timonium, MD: York Press, Inc.

Seewald, R., and Moodie, K. 1992. Electroacoustic considerations. In *FM Auditory Training Systems: Characteristics, Selection and Use*, ed. M. Ross. Timonium, MD: York Press, Inc.

Thibodeau, L., and Saucedo, K. 1991. Consistency of electroacoustic characteristics of FM systems. *Journal of Hearing and Research* 34:628–35.

Chapter • 6

Hardwire Personal Listening Systems

James J. Dempsey

THE ADA AND ASSISTIVE LISTENING DEVICES

The Americans with Disabilities Act (ADA) provides protection of rights for individuals with disabilities in both the public and private sectors. The ADA defines individuals with disabilities as those with conditions that substantially limit major life activities such as hearing, speaking, seeing, learning, breathing, performing manual tasks, working, and caring for oneself (Williams 1992a). Within this group of disabilities, hearing loss may be the most common condition. The number of Americans with significant hearing disabilities has been estimated to exceed 21 million (Williams 1992b). This figure represents approximately 50% of the population likely to be identified as disabled under the guidelines provided in the ADA (Gilmore 1992).

In an attempt to provide effective communication for individuals with significant hearing impairment, the ADA requires availability of assistive listening devices (ALDs) in certain settings. Under Title I, hearing aids are considered to be "personally prescribed devices" and, as such, are not the responsibility of the employer. The employer is,

Preparation of this chapter and the research by Dempsey and Ross cited within were supported by the National Institute on Disability and Rehabilitation Research, Grant No. H133E80019. This work is in the public domain.

however, required to provide some form of ALD as a "reasonable accommodation" in order to help a hearing-impaired individual perform essential job functions. Similarly, under Title III, facilities determined to be public accommodations are not required to provide hearing aids, but they are required to supply auxiliary aids or ALDs in order to achieve effective communication.

The ALDs provided under Titles I and III of the ADA are not mandated to be the most expensive or the most advanced technologies available. Employers or service providers may choose from the entire range of low-technology to high-technology types of equipment as long as the goal of effective communication is realized. Therefore, it is likely that these employers and service providers, in the interest of cost-effectiveness, will be looking for low-technology solutions to improve communication function with hearing-impaired individuals in order to comply with the ADA.

LOW-TECHNOLOGY AMPLIFICATION DEVICES

In recent years, many relatively inexpensive ALDs have become available to the general public. These simple personal amplifiers consist of a microphone, an amplifier, and some type of earphone transducer. The microphone on these devices is either contained in the unit itself or is directly connected to the unit via a cord or wire. Because of this direct connection between the components and the lack of a transmitted signal (FM or infrared), these devices are referred to as "hardwire" assistive listening devices. This type of hardwired listening system is illustrated in figure 1.

HARDWIRE ALDS VERSUS HEARING AIDS

These hardwire personal amplifiers should not be confused with hearing aids. As stated in the ADA, a hearing aid is considered to be a personally prescribed device. The federal Food and Drug Administration (FDA) does not classify these hardwire personal amplifiers as hearing aids. Therefore, they do not fall under the purview of FDA rules and regulations governing sales of hearing aid devices (USCA 1976; Federal Register 1977) and are not personally prescribed. Distinguishing between hardwire ALDs and hearing aids is often difficult because both types of devices amplify sound. However, the hardwire ALD generally amplifies less effectively than a personal hearing aid. Several manufacturers of hardwire ALDs include disclaimers in their advertising acknowledging that their devices are not hearing aids and

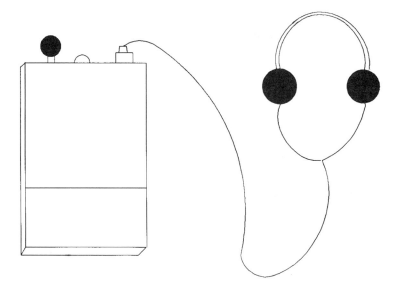

Figure 1.   A schematic representation of a hardwire ALD.

Figure 2.   The Listenaider. (Photograph courtesy of The Source.)

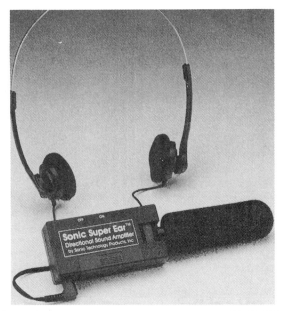

Figure 3.   The Sonic Super Ear. (Photograph courtesy of HARC Mercantile, Ltd.)

encouraging prospective purchasers to pursue a professional hearing examination. Figures 2 through 5 are examples of commonly employed hardwire ALDs.

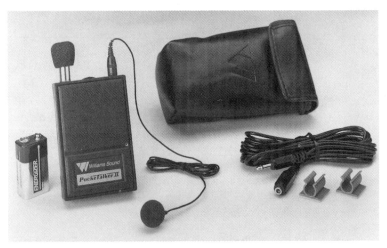

Figure 4.   Pocket Talker II. (Photograph courtesy of Williams Sound Corp.)

Figure 5.   The Whisper 2000. (Courtesy of The Source.)

USE OF HARDWIRE ALDS

The primary way to maximize communication with hearing-impaired individuals is to improve the signal-to-noise ratio (S/N) in a given environment. The S/N ratio is defined as the decibel level by which a primary signal (usually speech) exceeds the level of the background noise. Listeners with sensorineural hearing impairment require a S/N ratio of between +20 dB and +30 dB in order to achieve good word recognition scores. When employing a hardwire system, the S/N ratio is directly related to the distance of the sound source from the microphone of the amplification unit. As shown in figure 6, the closer the microphone is to the sound source, the greater the level of the primary signal and the more positive the S/N ratio. The listener's separation from the sound source is, therefore, limited only by the length of the cord connected to the microphone or the earphones (Compton 1989).

ROUTINE USES OF HARDWIRE ALDS

A hardwire ALD may, therefore, be particularly effective in certain situations in which a cord or wire is not problematic, such as watching television in the privacy of one's own home. Many hardwire ALDs

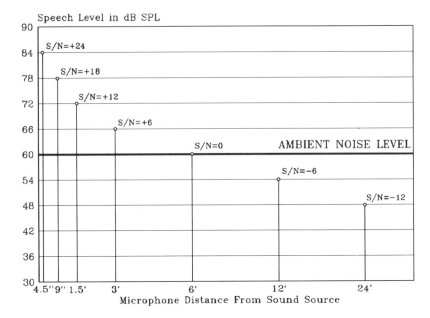

Figure 6.   Effects of microphone distance from the sound source on signal-to-noise ratio.

come with 20-foot extension cords that allow the microphone of the unit to be placed in close proximity to the loudspeaker of the television while the listener is comfortably seated a good distance away. This microphone placement provides a significant improvement in the S/N ratio for the hearing-impaired listener by increasing the intensity level of the primary signal. An added benefit of this arrangement is that the volume of the television loudspeaker can be maintained at a comfortable listening level for others in the home with hearing sensitivity within normal limits. Figure 7 demonstrates the use of a hardwire ALD for watching television at home.

Similar improvements in communicative effectiveness can be seen in any situation where the microphone can be placed in close proximity to the primary sound source. For example, when listening to music, the microphone of the hardwire ALD can be attached directly to the stereo system loudspeaker. The microphone of the hardwire ALD can also be attached to a public address system loudspeaker in a house of worship in order to improve communication during religious ceremonies. Conversational access can be improved in a restaurant or car by placing the microphone close to the individual who is speaking, as shown in figure 8.

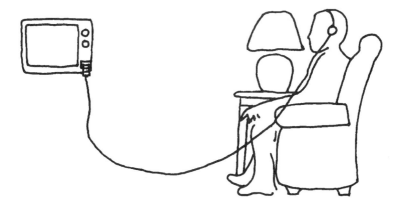

Figure 7.   An effective use of a hardwire ALD—watching television at home.

It should also be apparent that a hardwire ALD will not be of particular benefit in situations in which the sound source is a significant distance from the microphone of the unit. Figure 9 is an example of ineffective use of a hardwire ALD.

OCCASIONAL USES OF HARDWIRE ALDS

There are a number of specific circumstances in which hardwire ALDs may serve to facilitate effective communication. Doctors, lawyers, clergy, or other professionals may occasionally need to communicate with

Figure 8.   Use of a hardwire ALD in a restaurant and in an automobile. (Photograph courtesy of HARC Mercantile, Ltd.)

Figure 9.   An ineffective use of a hardwire ALD.

a hearing-impaired person. Figure 10 is an example of a physician communicating with a hearing-impaired patient through the use of a hardwire ALD. In a situation such as that depicted in figure 10, several advantages are provided by the hardwire ALD. The physician has direct control of the distance between the primary sound source (his mouth) and the microphone of the ALD. By holding the unit close to his mouth, the S/N ratio is significantly improved. An additional advantage provided by the hardwire ALD is that the unit can be used by any impaired listener because either "Walkman-type" earphones or earbuds are employed. There is no need for personal earmolds, as would be the case if one were to use either a body hearing aid or a behind-the-ear hearing aid to communicate in this circumstance.

USE OF HARDWIRE ALDS IN COMBINATION WITH PERSONAL HEARING AIDS

It is also possible to utilize a hardwire ALD in combination with a personal hearing aid. When used in this way, the hardwire ALD basically functions as an external microphone for the hearing aid. The advantage is provided because the hardwire ALD can be placed closer to the primary sound source, as compared with a traditional, internal hearing aid microphone located in or around the listener's ear. The result is an improvement in S/N ratio.

The hardwire ALD can be coupled to the personal hearing aid either acoustically or electromagnetically. Acoustical coupling is accomplished by placing the earphones of the hardwire ALD close to the microphone of the personal hearing aid. Although this arrangement can provide an improved S/N ratio, it is not commonly em-

Figure 10.   A physician communicating with a hearing-impaired patient via a hardwire ALD. (Photograph courtesy of Kevin E. Greagan, St. Peter's Hospital, Albany, NY.)

ployed because of the high probability of acoustic feedback.

The more common coupling technique for these types of devices is electromagnetic. In this case, the headphones or earbuds of the hardwire ALD are replaced with a neckloop telecoil coupler and the hearing aid telecoil ("T"-switch) is utilized. The advantages of this coupling technique include an improvement in S/N ratio when the microphone of the hardwire ALD is placed close to the sound source and an elimination of acoustic feedback. An example of this coupling arrangement is shown in figure 11.

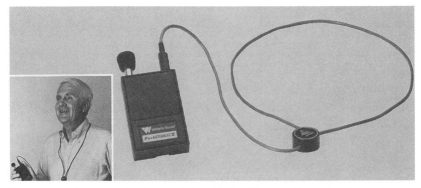

Figure 11.   An example of a hardwire ALD with a neckloop that is used to couple the device to a personal hearing aid electromagnetically. (Photograph courtesy of Williams Sound Corp.)

Although it is not considered a hardwire ALD, it must be noted that an external microphone can be coupled to a traditional hearing aid. Phonak, for example, manufactures an external microphone that can be coupled to a behind-the-ear hearing aid through the use of a direct audio input coupling arrangement. This external microphone can be placed closer to the sound source, resulting in improved signal-to-noise ratio.

USE OF HARDWIRE ALDS WITH SPECIFIC POPULATIONS

Successful use of hardwire ALDs has been reported with various specific populations. Erickson (1992) reports effective use of such devices with a cognitively impaired, mild-to-moderately hearing-impaired population. For this particular group, the relatively simple-to-use hardwire ALD serves as a substitute for the more complex personal hearing aid. The typical hardwire ALD is operated using a single volume dial and a set of earphones. Some individuals can manage such a system, whereas they cannot properly insert and manipulate a personal hearing aid. We should exercise extreme caution when making this substitution because an amplified signal provided by a personal hearing aid is likely to be far superior to that provided by a hardwire ALD.

Erber (1992) has reported successful use of hardwire ALDs with a mildly language-impaired adult population with normal peripheral hearing sensitivity. This group includes individuals suffering from aphasia, dementia, and attention deficit disorder who often have difficulty following spoken directions, particularly when such directions are given in background noise. Flexer and Savage (1992) suggest possible improvement in communication effectiveness with a language-impaired adult population using a hardwire ALD with a mild gain setting.

QUALITY OF THE HARDWIRE AMPLIFIED SIGNAL

Although it is possible to obtain improvement in the S/N ratio with all hardwire ALDs, many questions remain regarding the quality of the amplified signal they provide. Most manufacturers of hardwire ALDs do not provide electroacoustic specifications for their products. The limited specifications that are provided are often obtained with different transducers or employ coupling techniques with no applicability to the "real-world" listening situation.

In order to obtain descriptive information concerning these hardwire ALDs, Dempsey and Ross (1992) evaluated a number of these units electroacoustically. The devices were mounted on a Knowles Electronic Manikin for Acoustic Research (KEMAR) in an anechoic chamber. For each instrument, the manufacturer-provided transducer (i.e., earbud or Walkman-type phone) was coupled to KEMAR. Frequency response curves were obtained with a 60 dB input and each instrument set to approximate reference test gain position. The resulting frequency response curves for six commonly used hardwire ALDs are shown in figure 12.

In figure 12 we see wide variability in terms of frequency response across instruments. The amount of amplification provided varies significantly from one instrument to the next. Most of the frequency response curves have jagged peaks, and none of the instruments provides the flat, broad-band frequency response that we might—falsely—assume would be provided. These frequency responses would be considered very poor responses for a traditional hearing aid.

Another problem with these poor frequency responses is that little can be done to alter the output characteristics of these hardwire ALDs. In terms of electronics, these devices generally come with no trim pots or switches with which to adjust the frequency response or the maximum power output. In addition, little can be done through earmold acoustics to vary the responses because these units use such transducers as earbuds and Walkman-type earphones. They cannot, therefore, be varied to meet a particular individual's requirements when used alone.

As previously mentioned, these hardwire ALDs can also be coupled to personal hearing aids. Although significant gain can be expected, the resulting frequency response of such an arrangement is difficult to predict because it will be a product of the hardwire ALD itself, the properties of the personal neckloop, and the properties of the telecoil in the particular hearing aid.

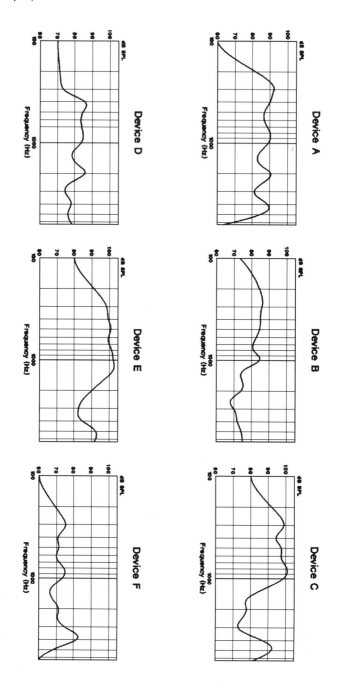

Figure 12.   Frequency response curves of six hardwire ALDs.

SUBJECTIVE EVALUATIONS OF HARDWIRE ALDS

Dempsey and Ross (1991) also evaluated each device subjectively in a group of 18 adults with confirmed sensorineural hearing loss who were members of Self-Help for Hard of Hearing People (SHHH). A range of severity of hearing loss was included in order to determine the relationship between degree of loss and preference for a particular device. An audiotape of a female speaker reading a passage was presented to each subject at a normal conversational level of 65 dB SPL (sound pressure level) while the subjects listened through the hardwire ALDs in random order. The subjects then rated speech understanding, quality of amplified speech, and ease of use of each instrument on a 5-point scale. Speech understanding was defined as the ability to recognize clearly the words produced by the speaker. Speech quality was defined in terms of the pleasantness or acceptability of the amplified speech. Ease of use was defined in terms of level of difficulty manipulating the controls and positioning the headphones or earbuds.

The results of the subjective evaluations of the hardwire ALDs revealed little variability for the ease of use category. Scores for this category were in the range of fours and fives, suggesting that each of the devices was very easy to use in the test environment. This positive rating may also be a result of the fact that an investigator was available during the test procedure to help each subject with any problems with equipment use. Due to this lack of variability, the ease of use category was dropped from all subsequent data analyses.

The data for the two remaining categories are shown as a function of hearing sensitivity as defined by pure-tone average (PTA) in figures 13 (speech understanding) and 14 (speech quality). In these two figures the solid bars represent a PTA between 26–45 dB HL (N=9) and the hatched bars represent a PTA between 46–70 dB HL (hearing level) (N=9).

In terms of speech understanding, it is evident from figure 13 that the most positive ratings were obtained for devices D and E by the group with the poorer PTA and device F by the group with the better PTA. Device E yielded fairly good scores from both groups, whereas devices B and C yielded fairly low scores from both groups.

Some similarities and differences are evident when comparing the speech quality ratings with the speech understanding ratings. As shown in figure 14, device F once again received the top ratings of the 26–45 dB HL PTA group. The quality ratings for the 46–70 dB HL PTA group were higher than the understanding ratings for devices B, C, and D, and most dramatically, for device A.

In looking for patterns regarding the ratings scores across hearing loss groups, we expected that there would be an underlying pref-

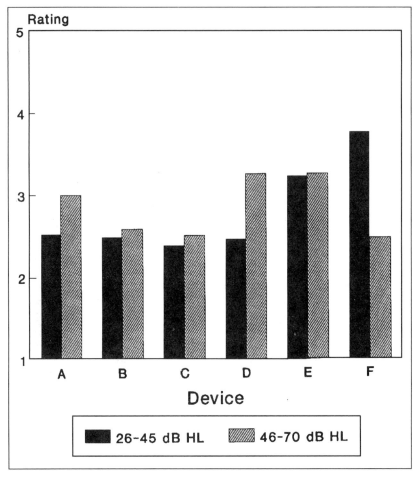

Figure 13.   Speech understanding ratings of the hardwire ALDs. The solid bars represent a group of subjects with a PTA between 26–45 dB HL (*N*=9), and the hatched bars represent a group of subjects with a PTA between 46–70 dB HL (*N*=9).

erence for instruments with greater gain. Figure 15 represents the average rating scores collapsed across subjects for both speech understanding and speech quality with the devices rank-ordered from most powerful to least powerful. Much variability remains, with only a slight trend toward lower ratings as amount of gain decreases.

A similar type of comparison is made in figure 16. The average rating scores for speech understanding and speech quality are presented with the devices rank-ordered from least expensive to most expensive. Interestingly, the scores obtained by the least and the most expensive

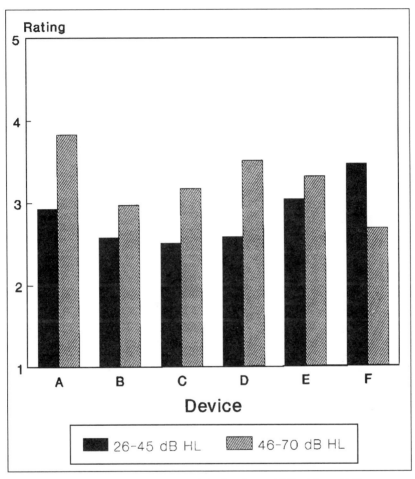

Figure 14.   Speech quality ratings of the hardwire ALDs. The solid bars represent a group of subjects with a PTA between 26–45 dB HL ($N=9$), and the hatched bars represent a group of subjects with a PTA between 46–70 dB HL ($N=9$).

devices are similar, and the highest rating is obtained by a moderately priced device.

The subjective analyses of the personal amplifiers varied greatly. None of the devices was rated statistically significantly higher than the others. Generally, the expected preference for power or gain was not evident in the subjective ratings. However, the importance of gain was emphasized anecdotally. When subjects gave devices low ratings, they often complained verbally about the lack of amplification being provided by the unit.

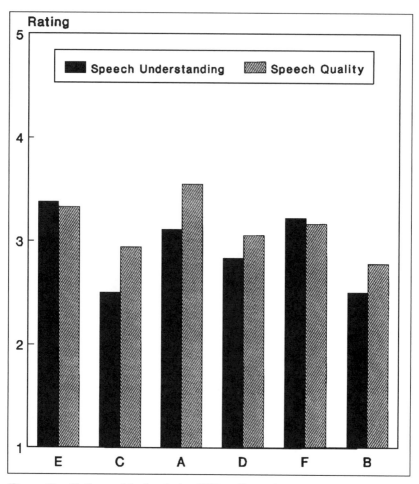

Figure 15. Ratings of the hardwire ALDs collapsed across subjects for both speech understanding and speech quality with the devices rank-ordered from most powerful to least powerful.

Interestingly, the group with the more severe hearing impairment (PTA=46–70 dB HL) did not prefer the more powerful instruments. This group demonstrated some preference for the moderately powerful instruments (devices A and D), while rating the most powerful instruments (devices C and E) slightly lower.

A cost/benefit analysis for this particular group of subjects would suggest that greater cost does not guarantee greater satisfaction. As shown in figure 16, the least and most expensive units produced similar results for the entire group of subjects. These data suggest that the minimal investment in device F is as likely to yield positive results as is greater investment in a more expensive unit. The most positive results were obtained with a moderately priced unit (device A).

HARDWIRE ALDS AND TRANSDUCER TYPE

The type of transducer employed with a hardwire ALD is a variable of significant importance. Dempsey and Ross (1991) analyzed and evaluated a number of hardwire ALDs employing the manufacturer-provided transducer (Walkman-type earphones versus earbuds) in order

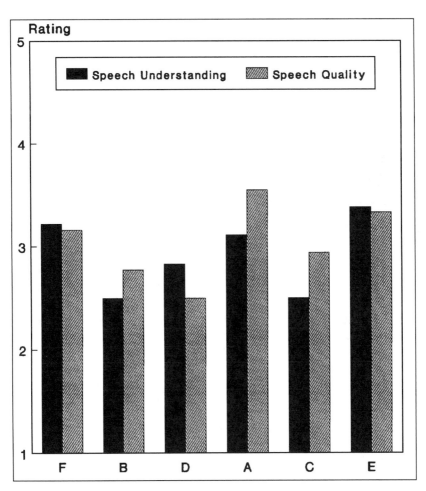

Figure 16.   Ratings of the hardwire ALDs collapsed across subjects for both speech understanding and speech quality with the devices rank-ordered from least expensive to most expensive.

to avoid impedance mismatches. This procedure also provided maximum face validity with regard to predicting outcomes for those who purchase these devices over the counter. However, a number of possible advantages exist for the earbuds versus the earphones. One primary advantage is improved comfort, particularly when wearing the units for long periods of time. In the Dempsey and Ross (1991) study, the majority of hardwire ALD users stated that they found the earbuds to be more comfortable than earphones.

In addition to providing greater comfort, the earbuds also provide a tighter seal in the ear, as compared with earphones, and a concomitant reduction in the leakage of amplified sound. Dempsey and Ross (1991) measured the amount of leakage with six different hardwire ALD systems by comparing electroacoustic characteristics obtained using a 2cc coupler with data obtained on KEMAR. As expected, due to the airtight seal obtained employing the HA-1 coupler, much higher output values were measured with the 2cc coupler versus KEMAR in each instance. Without exception, the differences between measurements were reduced when employing the earbuds versus the Walkman-type earphones. This reduction in leakage with the earbuds should translate into more effective use of the amplified energy, as well as reduced possibilities of acoustic feedback.

ADDITIONAL HARDWIRE ALDS

The discussion of hardwire ALDs in this chapter has been limited to devices costing under $100. More expensive personal amplifiers are available in today's market, such as the unit shown in figure 17. This unit contains potentiometers that allow for a variety of frequency response settings. It is possible that such a unit may be adjusted (within limits) to fit an individual listener's needs and, therefore, may provide a more appropriate amplified signal, as compared with other hardwire ALDs. Once again, the lack of available electroacoustic information makes this area worthy of further investigation.

ELECTROACOUSTIC PERFORMANCE STANDARDS

As mentioned previously, there are no published standards for assistive listening devices. Establishing standards in the future will be helpful in terms of selecting ALDs with appropriate output characteristics for a given listener. Standards would also be helpful in addressing the issue of quality control of these hardwire ALDs. At present, even if we have the equipment necessary to perform electroacoustic evaluations of hardwire ALDs, it is difficult to determine whether a given response is appropriate for a particular device unless we evalu-

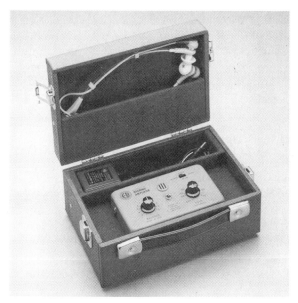

Figure 17.   The Model EB 140 Hearing Amplifier. (Photograph courtesy of Eckstein Bros., Inc.)

ate many units from the same manufacturer. Both the consumer and the dispenser will benefit from regulations requiring gain/frequency response information, a warranty of some sort, and some type of trial period for hardwire ALDs (Palmer 1992).

SUMMARY AND CONCLUSIONS

Hardwire ALDs should not be considered as substitutes for personally prescribed hearing aids. Hardwire ALDs should be considered as amplification devices with significant limitations, which also provide significant advantages to *certain* hearing-impaired individuals under *certain* circumstances. Table I summarizes the major appropriate applications of hardwire ALDs. As we look for ways to comply with the ADA, the number of hardwire ALDs in use is likely to increase substantially. Many manufacturers of hardwire ALDs have already included references to the ADA in their promotional literature. Therefore, it is necessary for the conscientious audiologist or hearing healthcare professional to be aware of the possible advantages and/or disadvantages provided by these low-technology assistive listening devices, when they are used alone or in combination with a personal hearing aid.

Table I.   Effective uses of hardwire ALDs

Used alone on a regular basis
   1.   To decrease distance between listener and sound source (i.e., watching TV)
   2.   As a precursor to a personal hearing aid, adaptation to amplification
   3.   For specific populations: cognitively impaired or head-injured persons, for example

Used alone on an occasional basis
   1.   To communicate with a hearing-impaired individual who does not use or does not have a personal hearing aid (i.e., doctor–patient)

Used in combination with a personal hearing aid
   1.   To serve as an external microphone and improve S/N ratio

SOURCES FOR HARDWIRE ASSISTIVE LISTENING DEVICES

HARC Mercantile, Ltd.
P.O. Box 3055
Kalamazoo, MI 49003-3055
1-800-445-9968

Williams Sound Corp.
10399 West 70th Street
Eden Prairie, MN 55344-3459
1-800-328-6190

The Source
National Hearing Aid Distributors`
145 Tremont Street
Boston, MA 02111
1-800-627-9930

Hal-Hen Co.
35-53 24th Street
P.O. Box 6077
Long Island City, NY 11106
718-392-6260

REFERENCES

Compton, C. 1989. *Assistive Devices: Doorways to Independence.* Washington, DC: Gallaudet University Press.

Dempsey, J., and Ross, M. 1991. Evaluation of low-technology assistive listening devices. Paper presented at the International Hearing Aid Conference: Signal Processing, Fitting and Efficacy, University of Iowa, Iowa City, Iowa.

Dempsey, J., and Ross, M. 1992. "Low-Technology" Assistive Listening Devices: Potential Benefits and Pitfalls. Miniseminar presented at A.G. Bell Association Annual Convention, San Diego, CA.

Erber, N. 1992. Effects of amplification on ease of conversation with hearing-impaired older adults. Unpublished paper.

Erickson, F. 1992. The Lexington Center. Personal communication.

Federal Food and Drug Administration. Federal food, drug, and cosmetic act, medical device amendments. May 28, 1976. USCA, Ch. 9, 360c (P.L. 94-295).

Flexer, C., and Savage, H. 1992. Using an ALD in speech-language assessment and training. *The Hearing Journal* 45:26–35.

Gilmore, R. 1992. Assistive listening systems: How ASHA members fit in. *Asha* 34:44–47.

Federal Register. 1977. *Hearing Aid Devices: Professional and Patient Labeling and Conditions for Sale.* Feb. 15, 1977. 42, (31):9286–96.

Palmer, C. 1992. Assistive devices in the audiology practice. *American Journal of Audiology* 1:37–51.

Williams, J. 1992a. What do you know? What do you do? *Asha* 34:54–61.

Williams, J. 1992b. Impact of the Americans with Disabilities Act on audiologists. *ADA Feedback* 3:7–9.

Chapter • 7

Television Amplification Devices

Dorinne S. Davis

How to hear the television is a major concern for a hard of hearing person. This concern is second only to the telephone when considering assistive listening devices. With passage of the Americans with Disabilities Act, the hard of hearing consumer, as well as those who deal with this population, should be aware of how improved television access can be attained.

Typically, we turn up the volume of the television when we want to hear "better." However, this may interfere with another person's ability to enjoy the show, whether that person is in the same room or the adjacent room. Often, this does not improve the ability to understand the message—it merely increases the volume. At other times, the hard of hearing person may hear only portions of what is being said. This difficulty can be overcome or minimized by using one of the many TV listening devices available.

TYPES OF DEVICES

There are four basic amplification devices for television: (1) infrared (IR); (2) FM; (3) hardwired; and (4) loop. Each has different characteristics.

These devices are known as assistive listening devices (ALDs) and they deliver an auditory signal to the ear in various ways. They were designed to improve the signal-to-noise ratio (S/N) for the listener. That means that the sound source, in this case the TV, is sent directly to the ear, eliminating or reducing the impact of distance, noise, or other interfering factors (Davis 1991).

The four types of assistive listening devices are either wireless or hardwired. The wireless ALDs use light rays, radio waves, or magnetic inductive energy to transmit sound. The hardwired ALDs use some form of direct electric connection to transmit the auditory signal.

Infrared Systems

Infrared systems are wireless systems that utilize invisible infrared light. The light beams are frequency and amplitude modulated.

They are produced by an array of light-emitting diodes (LEDs) (see Chapter 3, this volume). These LEDs are arrayed on an emitter panel, which typically is placed on top of a television set facing the listening area. The emitter is also known as the transmitter. The transmitter is plugged into an AC power source. Light rays emanating from the emitter panel are then radiated throughout the listening area, which in this case is the TV room. Because light rays do not travel through solid surfaces, sound transmission stays within the TV room (Compton 1989). A hearing-impaired person wears a receiver that "receives" the sound coming from the emitter/transmitter. The receiver uses a detector that demodulates or decodes the signal, amplifies it, and directs it to the listener's ear (Davis 1991).

The infrared beam must have an uninterrupted path from the transmitter to the emitter. That means that an object or person must not impede its transmission. Too much sunlight or artificial light will affect most systems currently available for TV use. Transmission is not affected by outside radio frequencies. Some systems can be connected with an additional microphone to enable conversation with others in the room. Of the currently available ALDs, IR is most effective for people with mild-to-moderately severe hearing impairment (Compton 1989).

Components of an IR system include a transmitter/emitter, a microphone, an amplifier, a receiver, and some way to couple the receiver to the ear, if it is not built in. There are a number of different types of receivers. The receivers are either worn on a person's body, or they are placed on a nearby table. Typically, the receivers are battery powered. One of the most common receivers is an under-the-chin variety, which fits into the ear with rubber tips and frequently is heavier than other receivers. Another type resembles a lanyard. It fits around the neck and uses either a neckloop, earbud, or pair of headphones. These coupling methods are discussed later in the chapter. Figures 1–4 display different TV IR systems.

Figure 1.   Components of an Audex infrared system with headphones.
(Photograph printed with permission by Audex.)

Typically, no tools are required to install an IR TV listening system. The user simply plugs the system into the audio output jack on the TV or places the microphone near the TV speaker. The first method typically cuts off sound for others in the room. The hard of hearing person will be the only one able to hear the sound. The second

Figure 2.   Person using an infrared listening system with a neckloop.
(Photograph printed with permission by Audex.)

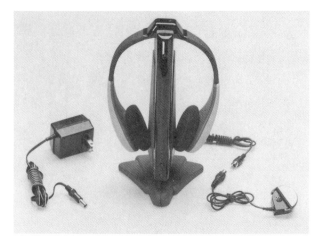

Figure 3. Components of Williams Sound's new infrared system. (Photograph printed with permission from Williams Sound Corp.)

method allows the TV sound to be heard by everyone in the room. The volume can be adjusted on the TV for those without hearing loss and can be adjusted on the receiver for those with hearing loss. A summary of the advantages and disadvantages of TV IR systems can be found in table I.

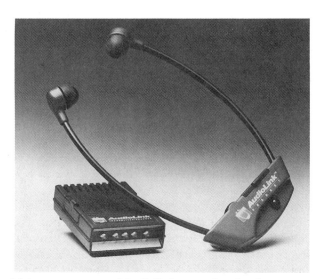

Figure 4. Major components of Audiolink, an under-the-chin receiver type. (Photograph printed with permission from National Captioning Institute.)

Table I.   Summary of Infrared System Characteristics

1.	Uses invisible light rays
2.	Wireless transmission
3.	Sound is confined to TV room
4.	Uses various types of receivers
5.	Can be affected by too much sunlight
6.	Can be affected by interrupting objects
7.	Is not affected by outside radio transmission
8.	Is effective for mild-to-moderately severe hearing losses
9.	Has AC powered transmitter/battery powered receiver
10.	Is somewhat portable
11.	Is easy to install
12.	Can allow hard of hearing person to hear TV with sound off
13.	Has excellent sound quality

FM Systems

FM systems are also wireless. An audio signal is modulated by frequency onto a carrier wave in a transmitter that is then sent to a receiver where it is demodulated and delivered to the ear directly (Davis 1991). In other words, the TV sound is changed into an electric signal and "broadcast" to an FM receiver.

The component parts include a transmitter, a microphone, a receiver, and some device to couple sound to the ear (see Chapter 5, this volume). The transmitter is attached to the microphone and is powered by a disposable or rechargeable battery. This makes the unit portable. If a battery charger is used, it is best to purchase the rechargeable battery recommended by the manufacturer for compatibility and length of battery drainage.

Some newer systems are built especially for television use. They offer stereophonic and monophonic sound and provide both high- and low-frequency emphasis, when necessary. Some units allow for separate right and left ear tone controls.

Installation is easy. The microphone attached to the transmitter is placed near the TV speaker. Some units can be connected to the TV with a direct plug-in connection through the audio output jack. The transmitter sends the audio signal to the receiver that is worn by the hard of hearing person. There is no wire connecting the transmitter and receiver. The receiver directs the sound to the ear via a neckloop, earbuds, headphone, silhouette, or direct audio input. The volume of transmission can be controlled at the TV or the FM receiver.

The Federal Communications Commission (FCC) has allocated the frequency bands of 72–76 MHz for auditory assistance devices (see Chapters 4 and 5, this volume). Both narrow-band and wide-band transmission are available. Previously there were 32 separate narrow-

Figure 5. Components of dB50 FM stereo listening system. Each ear can be balanced separately. (Photograph printed with permission of Chaparral Communications.)

band channels. These channels are spaced 50 KHz apart. The FCC recently amended its ruling (Federal Communications Commission, 92-163, April 7, 1992), and now there are 40 narrow-band channels available for transmission. Wide-band transmissions are spaced 200 KHz apart. Previously there were 8 channels. With the new FCC ruling, there are now 10 wide-band channels available. The availability of additional channels is especially helpful in heavily populated areas or in apartment complexes where one runs the risk of operating on a channel that a neighbor might also use.

Most TV FM systems can transmit their signal approximately 150 to 200 feet. The signal remains constant and does not vary when it meets an obstacle (Davis 1991). More powerful systems are available, but not usually needed for home TV use. Transmission radiates from the transmitter 360 degrees and travels through most media. This means that with typical home construction, an FM transmitter can be placed in the TV room, and the hard of hearing person will still be able to hear the TV program when he or she goes to the kitchen for a snack. An FM system can be used indoors and outdoors, and it does not depend on weather conditions. FM systems can be used by people without hearing impairment as well as those with profound hearing losses.

Various adaptations of FM use are available. One system allows the TV signal to be broadcast to a free-standing speaker that can be placed on a nearby table. A person can sit as close to the free-standing speaker as desired. Another application uses an FM system in conjunction with a loop system, which is described in detail later. A summary of the advantages and disadvantages of TV FM systems can be found in table II.

Hardwired Systems

Hardwired systems are so called because there is a hard wire connecting the unit to an earpiece. An amplifier might be in the unit near the sound source or at the ear itself.

The term *personal sound amplifier* has been used with one type of hardwired system (Davis 1991). This unit is often seen as a small hand-held unit that can be used alone or in combination with a hearing instrument. Such a unit usually is purchased with a set of Walkman-type headphones or an insert earbud. The unit consists of an amplifier unit, which has a microphone either built in or attached to it, and a connecting headset. Most units are battery operated.

An amplifying unit with a microphone is placed near the speaker of the television. Sound is picked up by the microphone, is changed into an electrical signal, the signal enters the receiver and is amplified, and the signal is changed back into sound at the earpiece. Most units come with an extension cord long enough to permit sitting between 9 to 15 feet from the TV (Compton 1989). This system can be used by people with any degree of hearing loss.

The sound quality of these units are often not as good as the infrared or FM systems, but many people prefer them over the most expensive systems. There are many inexpensive units on the market whose effectiveness is questionable because of their poor performance. The hard of hearing consumer should be wary of a price tag under $20.

Some personal sound amplifiers can connect to a TV set with a

Table II. Summary of FM System Characteristics

1. Wireless transmission
2. Most flexible system
3. Easy to install
4. Very portable
5. Battery operated
6. Person not confined to room for listening
7. Radio interference possible
8. Operates on radio waves
9. Inside and outside use possible
10. Effective for normal hearing to profound hearing losses
11. With adaptors, some systems can be jacked into TV

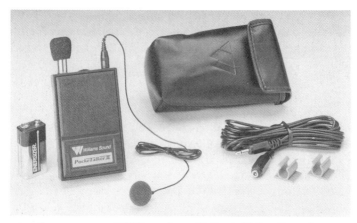

Figure 6.   Components of Williams Sound's PockeTalker Basic System. (Photograph printed with permission by Williams Sound Corp.)

direct connection cord. These cords must be purchased separately, either from the manufacturers or from electrical supply stores.

The other types of hardwired systems utilize a personal hearing aid. Many manufacturers have systems that connect a hearing aid to an external microphone, which can be placed near the TV, or to a jack that can be plugged directly into the audio output jack on the TV. The TV sound is directed to a hearing aid via direct audio input or a neck-

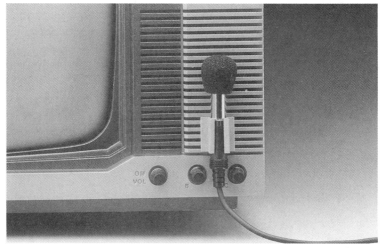

Figure 7.   Microphone placement near TV speaker. (Photograph printed with permission by Williams Sound Corp.)

Figure 8.   Using a hardwired system while watching TV. (Photograph printed with permission by Williams Sound Corp.)

loop. Both methods are discussed in detail later in this chapter. A summary of the advantages and disadvantages of hardwired TV systems
can be found in table III.

Loop Systems

Induction loop systems involve the electric principle of electromagnetic
induction (see Chapter 2, this volume). Sound is picked up by a microphone at the TV, converted into electric signals, amplified, and directed to a wire looped around an area or around the entire room. The
receiver is often a hearing aid that includes a "T" (or telecoil) setting.
The telecoil converts the electromagnetic signal from the wire loop
back into sound. Some hearing aids are not equipped with a telecoil.
When a hearing aid does not have a T-coil and a loop system is pre-

Table III.   Summary of Hardwired System Characteristics
1.   Very portable
2.   Easy to install
3.   Limited by length of wire
4.   Sound quality may not be good
5.   Battery operated
6.   Either hand-held units or directly connected to hearing aid

sent, a personal sound amplifier with a telecoil built into it can be used, or a special hearing aid that contains only the telecoil can be purchased. A person can sit anywhere inside or near the loop wire to pick up the TV sound. Signal strength is strongest when closest to the loop wire. The center of the looped area is often the weakest area for sound detection.

The loop system is unique because it is both a wireless and a wired system. It is wireless because the transmission of sound from the loop to the ear is received without an attached wire. It is a wired system in that the room is "wired" by a loop. Usually, however, an electromagnetic induction loop system is considered to be a wireless system (see figure 9).

Small loop systems are available for TV use. These loops are easy to install and are relatively inexpensive. The amplifier is AC powered and is connected to a television set by placing a microphone close to the speaker (see figure 7). A loop wire is connected to the back of an amplifier and is placed around the perimeter of the room or area. It can be placed under carpeting for safety or esthetic reasons. It is also possible to loop a single sitting area, such as a chair or couch, if pre-

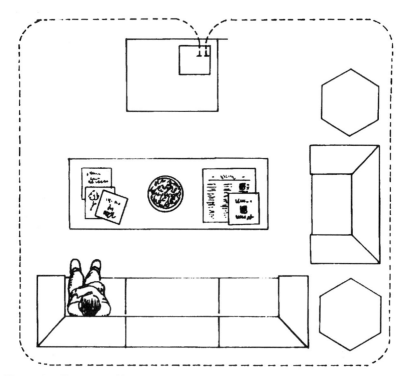

Figure 9.   Drawing of loop in use for TV.

ferred. A summary of the advantages and disadvantages of TV loop systems can be found in table IV.

Universal System

A new system called a universal listening system has been developed recently. The basic unit is a receiver that is the heart of the system. This receiver has a special directional microphone and works on a modular basis, which allows users to customize their listening needs. It can utilize either infrared, FM, telecoil, or direct audio input modules (see figure 10). This device can also work as a personal sound amplifier. By plugging in the appropriate module, it is possible to gain access to whatever system is already in place. For example, if a friend has an infrared TV listening system, a visitor can plug in the infrared module and become able to listen with his or her own universal listening system.

EVALUATION OF ALDS

The type of assistive listening device used depends on such factors as cost, availability, and personal listening requirements. User preference and choice have not been widely reported. However, one study evaluated the effectiveness of ALDs for older hearing-impaired individuals. Those surveyed expressed their preferences for ALDs in various settings. For home television use, they reported that they preferred infrared systems, hardwired systems were the second choice, and FM systems were the last (Hull 1988).

Alternative Systems

Other systems can be put together as a do-it-yourself project. The do-it-yourselfer needs to consider the importance of good quality of sound, correct volume settings, appropriate match for one's hearing level, and correct match for any electric components of the TV set.

Table IV. Summary of TV Loop System Characteristics

1. Works on magnetic induction
2. Use with a hearing aid with T-coil or special receiver unit
3. Limited portability
4. Installation: mild to medium difficulty
5. Variability of transmitting signal depends on proximity to loop wire
6. AC powered
7. Considered to be a wireless system

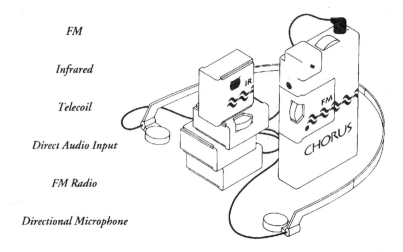

FM

Infrared

Telecoil

Direct Audio Input

FM Radio

Directional Microphone

Figure 10. Typical universal system with modules. (Printed with permission from Audiological Engineering Corp.)

Some such systems are:

1. A wired external speaker that can be plugged into an audio output jack or an earphone jack on a TV and placed near the listener.
2. An inexpensive FM system that allows a user to plug a transmitter directly into an audio output jack or an earphone jack on a TV, and transmit sound to a wireless speaker placed near the listener.
3. If the hearing aid has direct audio input (DAI), extension cords may be used to plug the DAI cord directly into an audio output jack or headphone jack.
4. A microphone, extension cord, and inexpensive amplifier can be combined to create a personal amplifier, as described earlier in this chapter.
5. Do-it-yourself loop systems have been described in various publications (SHHH 1989).
6. Some radios offer TV reception for VHF channels. A radio is placed near the listener and is tuned to the appropriate channel.

Headphones

One of the most typical attachments for commercial devices is a Walkman-style headphone. This system plugs into a receiver unit. The user simply puts on the headphones and listens. He or she hears the same sound coming through both headpieces.

Earbuds

The second most common attachments for commercial devices are insert earbuds. They come either monaurally (one ear) or binaurally (two ears), and they work similarly to headphones in that they jack into a receiver unit. Typically, earbuds are too large for children.

Neckloop

The neckloop is a smaller version of the area loop discussed earlier. It is worn around the neck and plugged into an output jack of an IR, FM, or hardwire receiver. The signal is electromagnetically transmitted from the loop to a telecoil placed in an ear-level receiver. This receiver amplifies the signal and converts it back into sound (Compton 1989).

The user can utilize his or her personal hearing aid as the receiver to conduct sound into the ear, as long as the hearing aid includes a T-setting. Some hearing aids have an "M/T" capability, combining both the microphone and telecoil settings. It is also possible to obtain a neckloop with an extra long cord for plugging directly into an audio output jack of a TV. This permits the person with the hearing aid to hear, but sound is cut off for others in the room. The volume can be controlled with either the volume control of the hearing aid or the TV volume control. Some TVs do not have this feature and may have other output jacks or none at all. This should be considered when purchasing a TV, if a neckloop is desired.

Silhouette Inductor

If a person wears a hearing aid with a telecoil, another way to access sound is with the silhouette inductor. Sound from the transmitter is sent via a cord that is plugged into the silhouette inductor. The silhouette inductor is placed next to a behind-the-ear hearing aid and changes the electrical signal into electromagnetic energy. This energy is picked up by the telecoil and converted into sound by the hearing aid.

The silhouette inductor can be used with infrared systems, FM systems, and hardwired systems. It is used only if the person wears a behind-the-ear hearing aid.

Direct Audio Input

Behind-the-ear hearing aids, and some in-the-ear hearing aids, can be equipped with a feature known as direct audio input. This allows a special coupling system, called a boot, to fit onto the end or side of a hearing aid. The boot connects to a cord that is jacked into the receiver of the unit.

The electric signal from a receiver is delivered directly to the hearing aid by a special electric connection on the hearing aid. It is possible to plug the direct audio input cord directly into a television audio output jack, but this cuts off sound for others in the room. It is also possible to connect a microphone to the end of a direct audio input cord and attach the microphone directly to a television speaker. Some hearing aid manufacturers have such a system available. Typically, direct audio input is used with FM systems, infrared systems, and some hardwired systems (see figure 11).

CONSIDERATIONS FOR SUCCESSFUL USE OF DEVICES

When people buy television amplifying devices, all too often their decision is based on two factors: cost and quality of sound. However, there is more to choosing the right device than these two factors.

The following items, beginning with cost and quality of sound, must be considered when purchasing a television amplifying device.

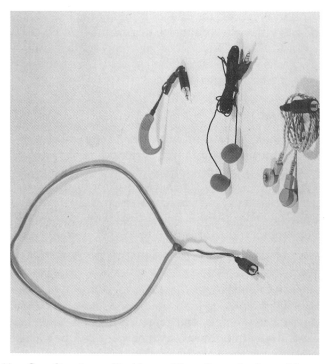

Figure 11.   Coupling items: Neckloop, silhouette, earbuds, and button ear-mold. (Photograph printed with permission from Audex.)

1. *Cost:* Cost is important for everyone. Devices can range from $25 to $1100. Some inexpensive devices on the market provide such a poor signal that the unit is not worth the cost. Be sure that you are getting your money's worth.
2. *Quality of Sound:* This factor is important for everyone. If a person watches only one hour of television a week and cannot afford an expensive device, then the quality of sound may not be a significant factor. However, if the person watches four hours of television daily and wants to enjoy it, then quality of sound is a major issue.

    Quality of sound is also a subjective issue. Some people prefer a simple amplifier to a more expensive infrared system, despite the superior sound quality of the latter device.
3. *Room size:* The size of a room may influence choice of the system. For example, a small room can be served with a hardwired or other advanced systems. A large room, however, may create problems when the length of cords have to be considered.
4. *Sunlight:* Presence of too much light will halt transmission of the infrared light rays. Makers of inexpensive infrared systems suggest turning off lamps when watching TV. Better infrared systems have solved this problem.
5. *Others watching TV:* For the person living alone or watching TV alone, any system will suffice. However, if others in the household watch TV with the hearing-impaired person, it is important to consider how the ALD unit will be connected to the television. If others are present, the microphone must be placed on the outside of the TV adjacent to the speakers. Using an audio output jack on the TV may cut out sound for others in the room.
6. *Portability:* If someone wants a system permanently attached to a TV, any system will work. However, if that person wants a system that can be taken on vacation or to visit other family members, then portability is important.
7. *Installation:* Technically oriented people have little difficulty understanding adaptive equipment. Others fear equipment and seek the least complex system. Some systems are simple to install—this should be a consideration when purchasing a device. It is preferable that a device be simple and be used, than be too complex and not be used.
8. *Comfort:* If the device is uncomfortable, it will be underused. Do users prefer to wear something dangling from the chin, to use Walkman-type headphones, or to make use of the convenience and proper fitting of their personal hearing aids?
9. *Expense of use:* Users need to determine if they want to purchase batteries for their listening system or use a totally AC powered

system. Occasionally headphones, earbuds, and neckloops must be replaced when wires or connections break in the cords. If units get dropped, what will be the cost of repairing the unit or system?

10. *Ease of use*: Placement of the volume control may be important. For example, does the user object to adjusting the volume control on the TV itself? Would he or she rather adjust the volume control on the body-worn unit? A summary of these ten points can be found in table V.

SUMMARY

There are four basic systems to consider when purchasing television amplification devices: infrared, FM, hardwired, and loop systems. Each has unique characteristics that must be considered.

The coupling system chosen should be evaluated for comfort, ease of use, and impact on hearing. All of these should be related to how well the system works in the environment in which it will be used.

Table V.   Considerations for Purchasing a TV ALD

Questions to consider before going to purchase	Questions to ask salespeople
1. Are other people going to watch TV with me?	1. Will the device turn sound off for those around me?
2. Do I want to wear something extra with or without my hearing aid in order to hear?	2. How comfortable will the headpiece be?
3. Can I watch the TV in a lighted room?	3. What devices work better under various conditions?
4. Do I want the best sound available for my hearing?	4. How expensive is a top quality unit?
5. Can I use the device on different types of TVs?	5. How portable is the device?
6. Can I install the device myself?	6. How technically complex is the device?
7. Is the size of the room important with the device I use?	7. How far can the device transmit?
8. How does the device attach to the TV?	8. Do I need an audio output jack?
9. Can I adjust the volume from anywhere I sit?	9. Where is the volume control setting?

With increased availability of listening devices, hard of hearing people should be able to enjoy television using their residual hearing. The message can be supplemented with captioning, but hearing should be the priority for the hard of hearing television viewer.

REFERENCES

Compton, C. L. 1989. *Assistive Devices: Doorways to Independence.* Washington DC: Gallaudet University.

Davis, D. S. 1991. Utilizing amplification devices in the regular classroom. *Hearing Instruments* 42(7):18–23.

Hull, R. H. 1988. Evaluation of ALDs by older adult listeners. *Hearing Instruments* 39(2):10–12.

SHHH. 1989. *Handyman Hints for Hard of Hearing Helps.* Bethesda, MD: SHHH.

Appendix: Manufacturer Product Availability

	Infrared	FM	Hardwire	Loop	Universal	Special Couple
Arkon	X					
Audex	X		X			
Audio Enhancement		X				X
Audiological Engineering Corp.					X	
Chaparral Communications		X				
National Captioning Institute	X					
One to One			X			
Oticon Corp.				X		X
Oval Window Audio				X		
Phonic Ear, Inc.		X				
Sennheiser Electronics Corp.	X	X				
Siemens Hearing Instruments Inc.		X				X
Sonovation, Inc.		X				
Telex		X				
Williams Sound Corp.	X	X	X			

Telecommunications

Chapter • 8

Telecommunications
Visual Technology

Diane L. Castle

There are very few jobs for the professionally employed deaf or hard of hearing person that do not require some form of telephone communication. During the 1970s and for many years before that, whenever deaf or severely hard of hearing persons were hired, it was assumed that those employees would not use the telephone. A supervisor would ask a hearing co-worker to manage the phone calls while employees with a hearing loss would assume some of the nontelephone tasks. The ability to use the telephone and participate in department meetings continues to be a real concern for employees with a hearing loss.

Today, options for telecommunication have expanded tremendously, but they may not be implemented in the workplace because neither the employer nor the employee is sufficiently educated about available technologies. Deaf employees, just as hearing employees, need to use several kinds of telecommunication technologies to accomplish their work.

The term *telecommunication* includes a wide range of technologies that transmit information from one location to another using telephone lines, modems, FM radio signals, satellites, or other systems.

This material was produced through an agreement between Rochester Institute of Technology and the U.S. Department of Education.

Over the years, deaf and hard of hearing people have been using text telephones (teletypewriters previously called TTYs/TDDs) along with relay services or amplified handsets for phone calls, while business and industry have been using computers, electronic mail, facsimile machines, pagers, and cordless and cellular phones. Deaf and hard of hearing employees recognize that many of these new technologies can enhance their job effectiveness because they provide a visual communication system or interface with a text telephone (TT). See figure 1.

In recent years, federal legislation has focused on telecommunications, as well as other areas of access for deaf and hard of hearing persons The impact of this legislation is stimulating development, modification, and application of technology to meet the needs of this special population. Legislative mandates are requiring compatibility of phones with hearing aid telecoils, text telephone access to federal and state agencies, and built-in closed captioned circuitry in 13-inch and larger television sets. The Americans with Disabilities Act (ADA) requires that telephone companies provide telecommunications relay services, 24 hours a day, seven days a week, at no additional charge, for all local and long distance calls within and between states. The law applies to state and local government employers, private employers regardless of the size of their business, and public accommodations in

Figure 1.   Text telephone. (Photograph used with permission from Ultratec, Inc.)

private facilities. Airports and other public places must provide access to TTs. Hotels, for example, must provide visual smoke alarms, TTs, and closed captioned televisions sets.

EXAMPLE A

> Hotels and motels are beginning to respond to the ADA. A deaf traveler, who checked into a hotel soon after the ADA went into effect, was given a room with a TT, a 19″ television set with a built-in closed captioned decoder and an alerting system that included a doorbell, smoke alarm, and TT signaler. After a while, the desk clerk called him on the TT to ask if everything was okay and if he wanted a wake-up call for the morning. Needless to say, the deaf traveler was delighted with the accommodations. Most of us would be unaware that, for a deaf person, these devices could change an unfriendly environment into comfort while away from home.

Although the ADA has opened up employment opportunities for disabled persons, it has placed on them the responsibility for suggesting which accommodations can maximize their potential in the work environment. Because of legislation, interviewers may not ask if applicants have a disability but they may ask applicants how they can perform the essential functions of the job with or without reasonable accommodations. Therefore, deaf or hard of hearing applicants should take the initiative to explain their use of the phone, their preferred methods for communicating and obtaining information in one-to-one and group settings, and any technologies, strategies, or other accommodations that would enhance their job performance. The employer will want to know the costs of these accommodations and how they will help to accomplish the job. At the same time, the interviewer may want to mention the kinds of telecommunication technology available to employees in the company. In many instances, it is a process of educating each other about the alternatives. Most technology useful to a deaf or hard of hearing employee is not expensive. Sources for such equipment are listed in the appendix to this chapter.

COMMUNICATION ABILITIES AND PREFERENCES

Deaf people vary greatly in their communication abilities; there is no relationship between hearing ability and intelligence. In the workplace, some deaf employees wearing hearing aids may benefit from using their residual hearing, some can use their speech, which may range from easily understood to nonintelligible, some may watch lips and facial expressions to speechread as a way to understand what is said, and some may rely on writing notes back and forth. Because deaf and many hard of hearing people depend on visual information to supplement their hearing ability, some may prefer to receive information or instructions in writing or in printed form to reduce communication misunderstandings. Many deaf people know sign language but most do not find it useful for fluent communication with hearing people in the workplace. However, some employers do hire sign language or oral interpreters to facilitate communication in selected meetings.

USING THE TELEPHONE

Many severely hard of hearing people who use their speech do have brief telephone conversations with family and friends but many feel technical, work-related information is more difficult to understand. In addition, some hearing people may speak very fast or not clearly or there may be background noise that interferes with understanding. When both people know each other's style of speaking and communicate regularly on the same topics, an employee may feel comfortable using the telephone. Typically, a telephone amplifier or hearing aid, or both, are used to make voices loud enough, and specific strategies make numbers, letters, words, and sentences clearer.

Telephone communication strategies, often used by telephone operators and airline agents, have been adopted by many hearing and hard of hearing people to ensure greater accuracy over the phone (Castle 1988). For example, ask for the spelling of a word and then repeat the misunderstood letter along with a familiar word that starts with the same letter: "Was that C as in Charles?" Or, if numbers are not clear, repeating them as individual digits or spelling the particular number can be helpful: "Was that 8-1-5, 8:15?" "Was that n-i-n-e?" Confirming important information before ending the phone call is the best way for both parties to be sure the message is understood.

TEXT TELEPHONES

Thirty years ago, Dr. Robert Weitbrecht, a deaf physicist, designed a modem that made it possible for deaf people to telephone others using a similar modem and teletype machine (TTY). Each time the phone rang, a lamp connected to the phone would signal the phone call. Since then, deaf and some hard of hearing people have been using non-speech text devices, similar to typewriter keyboards, with an acoustic coupler (modem) into which the telephone handset was placed.

Text telephones (TTs) provide immediate telephone interaction between two parties, each of whom has a TT. A TT is easy to use: turn it on, put the phone handset in the modem, dial the number, and begin typing when the other party answers. Each person can read the message on a narrow display screen as it is being typed and can take turns participating in the conversation. Some TTs also include a printer that provides a paper copy of the entire telephone conversation. As with other telephone-line transmission equipment, both parties must have compatible devices unless the call is made through a dual-party relay service. Hearing persons need to be aware of some basic information and etiquette related to using a TT or calling a TT user (Cagle and Cagle 1991).

Offices frequently install answering machines to record missed calls; however, most machines do not record TT tones because they are sensitive only to speech. At least two companies modify standard answering machines to accept TT messages; Heidico and MIRAC. When adapted for this purpose, the TT tones can be played back through a TT and the message can be read on the display or can be printed.

TT conversations take longer because of the built in slower transmission speed. To compensate, common English abbreviations are used frequently to save time. If a word is misspelled but can be understood within the context of the sentence, it need not be retyped. In addition, some punctuation, articles, or prepositions may be omitted if doing so does not interfere with meaning. The result is an efficient exchange of telegraphic but intelligible messages. An Example of a TT conversation follows.

HELLO MARY HERE GA [Go Ahead]

hi mary this is john how are u today q [question mark] ga

HI JOHN IM FINE HOW ARE YOU Q GA

im fine tx i just watnd xxx wanted to know if u r still meeting me at my offciexxx office tmw at three ga

YES ILL BE THERE RIGHT AT 3 PM SEE U LATER OK Q GA

ok great bye till then ga to sk (Stop Keying or good bye]

BYE SKSK

sksk

Before the ADA went into effect, a deaf employee requested a TT to facilitate phone calls to hearing clients using a relay service. The request was turned down; "If I buy you a TT, I'll have to buy other employees whatever they request."

In this situation, the employee recognized the need for a particular device but its usefulness may not have been explained clearly or sufficiently. Sometimes, seeing a demonstration of a TT call using the relay service adds to an understanding of the process. The employee may decide to bring a TT from home to use for a few weeks in order to demonstrate how this technology improves productivity and independence. Also, the employee might mention that without the use of a TT, co-workers have been taking time from their work to make these phone calls.

Today, many businesses focus on reducing costs rather than adding to their costs. Employers may be hesitant to purchase equipment without knowing its benefits. In some circumstances, an employee may need to buy and evaluate equipment during a 30-day trial period instead of expecting the employer to purchase it before demonstrating its cost effectiveness. After the evaluation, the employee is in a better position to discuss how this accommodation would be beneficial. When several pieces of equipment are needed by the employee, it may be reasonable for the employee and employer to share in the purchase of the equipment.

EXAMPLE B

A deaf employee, with a Master of Fine Arts in Computer Graphic Design, is working in a new business. He is the only deaf person in a group of eight people; all employees are working from their home offices to keep costs and overhead low.

The employee bought a TT and a facsimile (fax) machine for his home office. The company bought him a computer, a printer, and a second printer with wide paper for the TT. The employee uses the TT and relay service to call clients and they can call him through the relay service also. For his TT, he uses a separate printer with wide paper because it is easier to read than the narrow paper that comes with the TT. The TT has a built-in answering machine to record TT messages when he is not home. He has attached a small light to the telephone to let him know when the phone rings.

He uses his computer to send electronic mail to his boss and co-workers about the progress he is making on his projects. The fax is used to send artwork to his boss or his clients for their approval. The client or the boss may make changes and return the artwork to him. It saves the time of driving to meet the client, discussing the work, and driving back to the office.

ASCII—Baudot Modems

TTs and computers use different codes and different speeds for transmission; they are not compatible with each other. Baudot, used by TTs, transmits letters, numbers, and other characters at a slow speed

(45.45 bits per second). ASCII, used by computers, transmits characters at faster speeds (300, 1,200, 2,400, 9,600 bits per second). To communicate with each other, either the TT must have ASCII or the computer must be able to change from ASCII to Baudot. Some TTs are built with a computer chip and can communicate in ASCII code, but they must be reset to the appropriate specifications before making or receiving an ASCII call. Also, special modems and software can be purchased for ASCII terminals to switch the communication format for receiving TT calls.

Text Telephone Software

For deaf employees with a computer on their desks, TT software may be an obvious choice. Currently, there are several kinds of software (MICROFLIP and Futura TDD) and modems (Intele-Modem by Ultratec, Inc. and CM-4 by Phone-TTY) for IBM-compatible computers. For use with Macintosh computers, Dove Computer Corporation has developed the DoveFax TTY that combines a fax and TT modem with software. TT software does not accept telephone calls using speech or voice mail service. However, these programs do include many features found on modern telephones, including auto answer and remote retrieval.

How does the deaf employee use the computer for TT telephone calls? Each day, the computer and the modem for TT software are turned on. If a telephone call comes in, the employee can see the flashing message, "RINGING" on the screen while seated at the computer doing some word processing. The employee presses the correct keys on the computer keyboard to answer the call; meanwhile, the word-processing program is saved. When the call is finished, pressing the correct keys brings back the word processing. The ASCII-Baudot modem with software is comparable in price to a TT.

Recently, free "ASCII-TDD" software has been made available to encourage the use of ASCII among deaf, hard of hearing, and hearing owners of modems and computers (Telecommunication Assessment Program). Also, certain members of the Amateur Radio Research and Development Corporation (AMRAD) have developed conversion programs between ASCII and Baudot for various computers. They have placed the design information and software in the public domain. More information can be obtained by writing AMRAD or logging on to the Handicapped Educational Exchange (HEX) Bulletin Board System (301-593-7033), which is accessible to ASCII and TT. The Handicap News BBS also shares information about ASCII-Baudot communication programs and can be reached by TT at 300 baud (301-593-7033) or ASCII (301-593-7357).

At work, deaf or hard of hearing employees may prefer the two separate systems; a TT and a computer. Others may prefer a TT and a computer with a modem and TT software. Some people may want a pocket-size TT that also includes the ASCII code; this is especially useful for checking electronic mail and accessing computers when out of the office. Both the AT&T Accessible Communications Product Center and Ultratec, Inc. sell these small TTs.

Each year, Telecommunications for the Deaf, Inc. (TDI) publishes an international directory that lists TT users including organizations, businesses, services, and manufacturers of TTs; this organization serves as a resource for information about TTs and ASCII-Baudot modems. TDI has produced a captioned videotape about TTs and how to use them including etiquette, abbreviations, and turn-taking. To help new TT users become more proficient, the Deafworks Company has designed a TT trainer device that allows two or more people to practice making calls to each other without tying up phone lines. The training device comes with its own telephone handset for each TT and a standard power outlet.

Pay Telephone TDD

TT-equipped pay telephones make many public facilities accessible, including hospitals, airports, schools, hotels, and shopping malls. The Ultratec Pay Phone TDD sets the TT into a metal, vandal-resistant drawer that opens in response to receiving TT characters from the other party and closes when not in use (figure 2). Hearing persons can use the same pay phone without any interference by the TT. Locations of public TT pay phones are identified by the international symbol (figure 3).

AT&T Public Telephone 2000

Another kind of telephone seen in airports and hotels is designed for all travelers, including persons who may be deaf or hard of hearing. The AT&T Public Phone 2000 has a keyboard for use as a TT or a dataport to connect a laptop computer, portable fax machine, or TT, and an amplified handset that is compatible with a hearing aid. This telephone is typical of a design trend toward technology that can be used by all people.

TELECOMMUNICATION RELAY SERVICES

For many years, deaf and hearing people were unable to call each other unless both parties had TTs. Today, using a telecommunications

Figure 2.   Ultratec pay telephone TDD. (Photograph used with permission from Ultratec, Inc.)

relay service, deaf, hard of hearing, speech-impaired, and hearing peo-ple can telephone each other. Relay services have two separate tele-phone numbers: one is used by hearing people and the other by TT users. Both numbers should be listed in the front pages of the tele-phone directory, or they can be obtained through directory assistance.

The relay service transmits conversation between a person using a TT and a person using a voice telephone. Three people are involved: the person who uses a TT, the person who does not have a TT, and a communication assistant (CA) or agent who works for the relay ser-vice (figure 4). The person using the TT types; the CA speaks every-thing that is typewritten. When GA (go ahead) is typed, the other per-son can begin to speak. Everything that is spoken will be typed by the CA to the TT user. Each party communicates directly to the other per-son as if the CA were not on the line. All information is confidential and no records of the conversations are kept. Specific knowledge and skills must be learned by the CA and a code of confidentiality is as-sumed regarding all telephone calls. Some relay services are compatible

Figure 3.   International TT/TDD/TTY symbol.

with ASCII and provide special instructions for setting up the computer software.

Voice CarryOver

Many state relay services have introduced a feature called Voice CarryOver (VCO). VCO allows a deaf or hard of hearing person, who prefers to speak, to talk directly to a hearing person instead of typing on the TT. The CA only types what is spoken by the other person.

The deaf person must inform the CA that VCO will be used. Then, the CA will set up the appropriate connections and indicate when the deaf person can begin to speak. The telephone handset is moved into and out of the TT modem, depending on whether the person is talking or reading what is being said. Another way to set up the TT for VCO involves using a single to double line phone adapter plug. The direct connect TT and the telephone base are connected to the

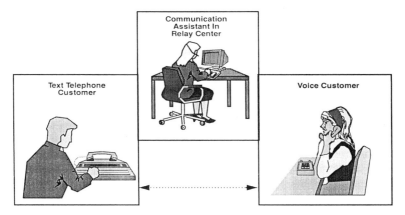

Figure 4. Telecommunication relay service. (Illustrations used with permission of *SHHH Journal*, March/April 1993 issue.)

adapter, which is plugged into a wall outlet. The handset is used normally instead of putting it into the modem.

Deaf and hard of hearing users, whose speech and language can be understood on the telephone, find VCO a more accurate way to convey their personality, feelings, and attitudes than by using the CA. VCO calls go much faster than the standard relay process.

EXAMPLE C

A deaf accountant describes her use of several telecommunications technologies during the workday. When clients call her using the relay service, the accountant uses her own speech to talk directly to the client. She reads what the client says on the display of the TT. It is critical that the client understand all the information clearly. To be sure there is no confusion, at the end of the phone call the accountant will fax a brief summary of the discussion, including any calculations or figures. The fax serves as a record of the call for both the client and the accountant.

COMPUTERS

All over the world, people in business, industry, and education can communicate with each other and share data and files without leaving their desk. Special computer software programs link people together and eliminate the inconvenience or expense of face-to-face meetings. In a growing multicultural and multiaccented population where spoken English may have been learned as a second language, using the computer for text communication may improve comprehension. For many deaf and hard of hearing employees, the text is less ambiguous than speechreading or listening.

Many deaf and hard of hearing people are comfortable using computers because of their experience with TTs and text conversations. The computer provides opportunities to be fully involved with colleagues and to participate in team efforts by sharing ideas, analyzing problems, displaying work, and contributing to the decision-making process. Unfortunately, some hearing employees may be reluctant to use computers, excusing themselves by saying that they type too slowly or they may not express their ideas clearly; they would prefer face-to-face communication.

Electronic Mail

Electronic mail is similar to sending a letter and waiting for a response; it is designed primarily for sending short messages, memos, or files to individuals or groups of individuals without using paper. Both the sender and receiver have a permanent copy of the mail until they decide to discard it. Deaf and hard of hearing employees benefit from using electronic mail, especially when others in their departments use it, too.

Electronic mail is cost-effective and confidential because each person has a password. Many users enjoy the freedom of reading and answering messages at their convenience; telephone calls and telephone tag are practically eliminated. Messages can be sent to people in the same building or at distant locations; the same message can be sent to more than one person at the same time. Engineers can send data to each other or test a computer program without a face-to-face meeting. Meeting agenda can be sent and minutes distributed. Saving time is another asset of electronic mail.

Local area networks (LAN) provide the necessary connections for computers to send and receive electronic mail from users within or outside specific businesses or educational facilities. If the company does not have a LAN, access to electronic mail is available to subscribers of the same information services, for example, America Online, CompuServe, Prodigy.

There are, however, some disadvantages to electronic mail: (1) it is not interactive as telephone conversations are; (2) all persons who wish to send and receive messages must be part of a LAN and have the appropriate software; and, (3) not everyone uses electronic mail, and those who do may not check it daily.

Conferencing

Conferencing is similar to using a bulletin board; information can be pinned to a bulletin board and shared until it is discarded. Computer conferencing allows discussion group members to read and respond to different topics stored on the computer system. Input to the discussion topics can occur at different times and from various locations. The information remains under the topic until it is deleted.

Conferencing is a way for deaf and hard of hearing employees to participate in a group or team meeting about specific issues without being placed at an auditory or visual disadvantage. In face-to-face meetings, the swift interaction of ideas being shared and reacted to may inhibit the participation of some employees, not only those with a hearing loss. Conferencing is a productive way to gather information that may lead to consensus. One disadvantage is that conferencing is not interactive on a real-time basis.

Interactive Communication

Interactive communication is similar to telephone or TT calls (figure 5). It is temporary with no record of the information, unless the telephone is connected to a tape recorder or the TT has paper copy. Interactive communication allows two or more people to have a real-time conversation. The information is displayed on the screen only as long as it is needed, and it is not automatically saved. Data or graphics displayed on the computer screen can be shared and worked on together. Individuals can communicate, using their computers, as if they were sitting together in a meeting room or next to each other at a drafting board. When information is presented on the screen, all participants can see the same material at the same time. Involvement in the discussion and the outcome is shared by all. Not only does the computer offer a practical solution for reducing the frustration of communication if speech is unclear or not used, but it encourages turn-taking.

Projecting the Computer Screen

Information displayed on a computer screen can be enlarged for the benefit of persons attending a meeting. Using a video interface, an overhead projector, and a liquid crystal display (LCD) projection panel, the image is projected for all to see. Another approach is to use a TV monitor and connect a converter box and appropriate cables between the computer and the television set. If a secretary or the presenter types in the content of the discussions, each attendee would have access to all information and it could be retained for those who

Figure 5. Interactive communication. (Photograph by Mark Benjamin.)

were absent. This technology can be extremely helpful to deaf and hard of hearing employees attending meetings.

EXAMPLE D

Recently, the graphic designer referred to earlier moved from his home office to work in a downtown office building. He is still the only deaf employee in a small company. All employees use their computer network to log in their progress on each phase of a project. Group meetings are held in a conference room equipped with a computer. Employees reporting on a project use the computer connected to an overhead projector; on top of the projector is an LCD panel. This panel makes it possible to show their work in color and enlarged. The employee who is presenting information about the project types the information into the computer. It is projected, enlarged, and easily read by everyone in the room. These electronic meeting notes are sent to each employee's computer file. The employee can decide to retain or discard the file, or to send it to the laser printer. If an employee leaves the room or misses a meeting, the electronic notes serve as a record of the discussion.

The director of the company indicated that these procedures were not used because of the deaf employee, but for efficiency and to ensure that everyone was informed of current projects.

The PhoneCommunicator

A unique software program has been designed to translate a typed message from a deaf person into synthetic speech and transmit it to a hearing person listening over a telephone. The hearing person, using a touch-tone telephone keypad spells out words, using the numbers and letters; the tones are converted into words that appear on the computer screen. The PhoneCommunicator has an internal modem, can use ASCII or Baudot, and can run on IBM compatible computers.

There are two disadvantages to this system; the software is expensive and spelling words on a touch-tone keypad is time consuming. This program was innovative when first introduced because statewide relay services had limited availability; however, the ADA now ensures full accessibility for relay services throughout the United States.

FACSIMILE EQUIPMENT

Facsimile or fax machines send copies of documents, either text or graphics over telephone lines. Using this technology can save time on the telephone and ensure accuracy because specific information or documentation can be faxed to avoid spoken misunderstandings or to verify information. For certain deaf and hard of hearing employees, reading a faxed message also reduces their dependence on comprehension through listening.

Fax machines are easy to use: merely dial the telephone number, insert the paper to be sent, and push a button to start the transmission; within seconds a copy of the paper is received. Many computers can accommodate an internal fax board and eliminate the need for a separate piece of equipment. A dual plug can be used to share a telephone line between fax and phone calls. If fax messages are sent and received frequently, it may be preferable to install separate phone lines for each device so that fax and telephone calls can be received simultaneously. Fax communication may not be private if the document is received in a central location before it is sent to the designated person.

PAGERS

Pagers are one of the most efficient ways of contacting a person who moves from one location to another. An individual who wears a pager is alerted to incoming messages by vibration (silent alert) or tone. Early models transmitted only the tone or a voice message; later, digital and alphanumeric (text) displays became available. The digital pager only shows numbers; for example, a telephone number to call. Text pagers can display a message and a telephone number, as well as the name of the person sending the page. The newest text pagers can show four lines of 20 characters and scroll to show the rest of the message (figure 6). Voice mail can be teamed with a paging service to notify people immediately when a message is received. Pagers are becoming smaller, thinner, and lighter; some are the size of a credit card.

Most paging is used to give information instead of requesting a follow-up telephone call. Text pagers, in the vibration mode, can provide easy visual communication for the deaf or hard of hearing employee.

There are four different methods of sending a message to a text pager. The first method is with the assistance of the message center dispatch operator. The caller dials the message center and tells the

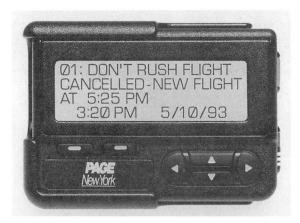

Figure 6.   Alphanumeric (text) pager. (Photograph by Mark Benjamin.)

message to the operator who transmits it to the pager using a specially designed keyboard.

The second method requires the use of a special keyboard also; some businesses have purchased the necessary equipment for paging their employees. Anyone who wants to send a page would call the company's paging operator. The operator would type the message on the keyboard, dial the pager number, and send the information directly to the pager.

The third method permits TT users to page someone without using a relay service to contact the message center dispatch operator. TT users can type a text message directly to the pager if the paging company is using a special TT interface.

The fourth method permits computer users to page someone directly without using a specialized keyboard. There are software versions for IBM-compatible, Macintosh, and other computers. Friends and family members who have access to computers and the software can page the employee also.

As deaf and hard of hearing employees assume more job responsibilities and spend more time away from their desks, text pagers will keep them in touch with co-workers and their supervisors. The palm-sized pagers that vibrate as well as flash a light can enhance productivity by reducing the time spent on telephone calls. When telephone contact is needed, cellular phones can be used with TTs. Pagers used in the vibration mode must have batteries changed more frequently.

EXAMPLE E

> An employee with a Bachelor's degree in Civil Technology works as a construction inspector for the county. He has been on the job for eight months and is the only severely hard of hearing person in the five-person department. He travels to three or more sites each day to inspect the job. The employer provided a vibrating pager and a TT that he used with a relay service to call his office. When he was out in the field, he was signaled by the vibrating pager. Then, he had to drive around to find a pay telephone and call his office using the portable TT and the relay service. His call to the office told him where he must go for the next inspection. This happened several times each day.

> Recently, the employee received a vibrating text pager. All the information he needed was shown on the pager and eliminated use of the TT, the relay service, and frequent calls to the office.

VOICE MAIL

Many businesses are replacing the switchboard operator, personal secretary, and the answering machines with a computer-based system called automated attendant or voice mail. This technology answers telephone calls with a recorded voice that asks you to press a number on the touchtone keypad to route the call to the correct extension or "mailbox" for the person or service you are requesting. Until recently, the TT user could not access information or leave a message using voice mail. Now, there are a number of companies that provide voice messaging services for TT users including Dirad, Magellan, Rolm, AEI/TeleSonic, and Tower. Depending on the company, the voice messaging system may use two separate telephone lines or integrate voice and TT on the same telephone line. Also, some companies provide the option of receiving fax messages on the same system.

An example of how the system works for TT mail is shown in figure 7. The numbers across the top represent the choices seen by the caller on the TT display. Pressing 1 on the touchtone phone keypad gives the caller access to that information or service. Pressing 2 on the keypad gives access to different information, and so on. Frequently, the explanation of these choices moves ahead rapidly and if there are more than a few selections, it is difficult to remember them. Being able to see the choices on the TT display is helpful.

Because the pace of voice mail is so fast, some state relay services have been unwilling to transmit this information to the TT caller because it could require several calls before all the choices could be typed. Now that the technology for TT mail is available, the deaf or hard of hearing caller can be more independent.

VIDEOPHONES

The idea of a picture telephone has always generated enthusiastic support from deaf and hard of hearing people because the addition of facial expressions, lip movements, or signing could make communication easier. Early demonstrations at the 1964 World's Fair piqued the interest of viewers but it never became a viable choice because of expense and video bandwidth requirement. The increasing use of fiber

Press 1	Press 2	Press 3	Press 4	Press 5
For General Info	For Info About New Programs	To Order	For Operator With TT	Repeat

Figure 7. Example of TT mail.

optics over the past few years has promoted a resurgence of interest in video communication systems by business, education, and government organizations.

Recently, small personal-use videotelephones that use regular telephone lines have become available but the picture moves at only 2 to 10 frames per second and appears stilted. This speed is not sufficient for sign language communication or for speechreading. Compared with home television pictures at 30 frames per second or videoconferencing systems used by business at 15 frames per second, picture telephones are not a communication panacea yet.

Another approach to picture telephone communication involves connecting a camera and closed-circuit monitor together to create a video intercom. This arrangement is similar to security monitoring devices in some buildings. This less expensive approach may be useful for employees who use speechreading, some residual hearing, or signing.

ASCII-BAUDOT PHONES

The recently introduced Philips Screen Telephone is about the same size as a regular feature telephone, and it offers both Baudot and ASCII, a 5-inch, 16-line LCD screen, a slide-out keyboard for text communications, and a printer port (figure 8). For those able to speak but unable to hear, with one touch of a button, it is possible to change between voice and text during the conversation. The design of the Philips Screen Telephone makes it possible to add new software applications as they become available. However, it does not include a video component.

MULTIMEDIA TELEPHONES

It is hard to believe that technology is changing so rapidly. The device of the future is described as multimedia: part telephone, part television, and mostly computer. With a multimedia device, the user could write on an electronic pad and fax it. Pushing a button could turn the electronic pad into a "touch screen" that guides the user through a series of choices for home banking and other services. Pushing another button could turn the touch screen into a video screen where you can see and talk to callers. There is no doubt that multimedia telephones will offer tremendous benefits to deaf, hard of hearing, and speech-disabled persons.

INTEGRATED SERVICES DIGITAL NETWORK

Integrated Services Digital Network (ISDN) follows a set of interna-

Figure 8.   Philips Screen Telephone. (Photograph used with permission from Philips Home Service.)

tional standards. Companies that adhere to these standards can communicate sound, computer data, or an electronic image to any other ISDN user in the world. Now, most companies in the United States send voice by telephone, data through the computer, and pictures through television on separate transmission lines to be received by individual telephone, computer, and television equipment. ISDN eliminates the need for separate lines and equipment by transmitting sound, computer data, and electronic images through one line and one display unit. As copper phone lines are replaced by fiber optics, deaf and hard of hearing employees will be able to maximize their use of visual information for communication.

FINAL COMMENTS

Increasingly, the technologies used by deaf and hard of hearing employees will be the same as those used by their colleagues. As employers upgrade their telephone and computer technologies, there will be more opportunities for deaf and hard of hearing employees to demonstrate their technical, collaborative, and leadership skills. There is no single solution that will enhance telecommunication accessibility in the workplace; the options will depend on the individual, the job, and the attitudes of supervisors and co-workers.

REFERENCES

Cagle, S. J., and Cagle, K. M. 1991. *GA and SK Etiquette: Guidelines for Telecommunications in the Deaf Community*. ($8.95 from Telecommunications for the Deaf, Inc., 8719 Colesville Road, Suite 300, Silver Spring, MD 20910.)

Castle, D. L. 1988. *Telephone Strategies: A Technical and Practical Guide for Hard-of-Hearing People*. ($6.00 from Self Help for Hard of Hearing People, Inc., 7800 Wisconsin Avenue, Bethesda, MD 20814.)

Castle, D. L. 1993. *New Solutions to Old Problems: Telecommunications for Deaf and Hard-of-Hearing Employees*. (Single copies free from Rochester Institute of Technology, NTID, 52 Memorial Drive, Rochester, NY 14623-5604.)

Using Your TTY. 1993. [Captioned videotape.] ($34.95 from Telecommunications for the Deaf, Inc., 8719 Colesville Road, Suite 300, Silver Spring, MD 20910.)

APPENDIX: RESOURCES

AEI/TeleSonic
120 Admiral Cochrane Drive
Annapolis, MD 21401
AMRAD
P.O. Drawer 6148
McLean, VA 22106-6148
AT&T Accessible Communications
Product Center
(formerly AT&T Special Needs Center)
5 Woodhollow Road, Rm 1-I-19
Parsippany, NJ 07054
1-800-233-1222 (Voice)
1-800-833-3232 (TT)
Dirad Technologies, Inc.
14 Computer Drive East
Albany, NY 12205
Dove Computer Corporation
1200 North 23rd Street
Wilmington, NC 28405
Futura Wave Communications
7209 Cipriano Springs Drive
Lanham, MD 20706
Heidico
561 Keystone Ave., #296
Reno, NV 89503

IBM PhoneCommunicator
IBM Independence Series
Information Center
P.O. Box 1328
Boca Raton, FL 34429
Magellan Communications, Inc.
1292 Hammerwood Ave.
Sunnyvale, CA 94089
MIRAC
545 Route 62
Winchester, OH 45697
MICROFLIP, Inc.
11211 Petsworth Lane
Glenn Dale, MD 20769
Philips Home Services
8 New England Executive Park
Third Floor
Burlington, MA 01803
Phone-TTY, Inc.
202 Lexington Avenue
Hackensack, NJ 07601
Rolm
4900 Old Ironsides Drive
Santa Clara, CA 95052

Telecommunications for the Deaf, Inc.
  8719 Colesville Road, Suite 300
  Silver Spring, MD 20910
Tele-Consumer Hotline
  1910 K St., N.W.
  Suite 610
  Washington, DC 20006
Telecommunication Assessment
  Program
  Gallaudet University
  800 Florida Avenue, N.E.
  MSSD 200
  Washington, DC 20002

The Deafworks Company
  1106 South State Street
  Suite 17
  Provo, UT 84606-6347
Tower Communications
  Attn: Keith Bass
  5820 Wilshire Blvd.
  Suite 503
  Los Angeles, CA 90036
Ultratec, Inc.
  450 Science Drive
  Madison, WI 53711

Chapter • 9

Telecommunications
Acoustic Technology

Alice E. Holmes

Trying to understand speech transmitted though a telephone may create unique and difficult listening problems for many persons with hearing impairment. These situations occur because of the need to rely totally on impaired auditory abilities with no visual cues and because of the limited electroacoustic characteristics of the telephone system.

In 1992, Kepler, Terry, and Sweetman reported on a survey of 104 adults with hearing impairment. Seventy-five percent of their respondents found telephone listening to be "somewhat"-to-"extremely" difficult. Sixty-nine percent stated that their hearing impairment discouraged their use of the telephone. Fifty-one percent indicated they sometimes avoided telephone use, whereas nineteen percent frequently avoided telephones. Respondents to this survey were members of the Self Help for Hard of Hearing People, Inc. (SHHH). The results revealed the extent of difficulty telephone use presents to a relatively well-informed segment of the hearing-impaired population. These dismal statistics on telephone listening problems are probably worse in the general hearing-impaired population.

ELECTROACOUSTIC CHARACTERISTICS OF THE TELEPHONE

In order to understand the problems associated with telephone listening encountered by individuals with hearing loss, the electroacoustic characteristics of the telephone system must be known. The acoustic output of a telephone is dictated by the frequency and line characteristics of the entire telephone system. The majority of telephones in use today employ a light-weight, granular, carbon-type microphone for the transmitter and an electromagnetic receiver (Erber 1985; Fike and Friend 1983). Transmitter output varies as a function of frequency. That is, output increases as frequency increases from approximately 250 to 2500 Hz followed by a precipitous drop in output above 3500 Hz. Therefore, the transmitter is the primary frequency limiting device in a telephone set (Ingles and Tuffnell 1951). Although the transmitter has less than ideal electroacoustic characteristics, it is still used for reasons of economy and because its limited frequency range is still acceptable for transmission of most speech sounds (Richards 1973). Persons with hearing loss, however, may find this limited response to be less than adequate.

Because the frequency response of a telephone is not uniform, it is difficult to specify the sound pressure level (SPL) transduced through the telephone. However, Bell Northern Research (Stoker 1982) has reported that the average output of a normal conversation is about 86 dB SPL at 1000 Hz (figure 1).

In addition to the limited response of the telephone signal itself, various types of interference or line noise may impair telephone speech perception. Intermittency or static can be caused by any part of

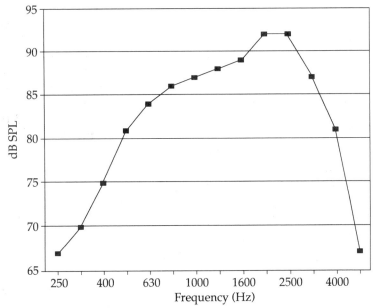

Figure 1.   Frequency response of a typical telephone.

the telephone system (Erber 1985; Fike and Friend 1983).

Another potential source of interference is the telephone sidetone system in a telephone handset. The sidetone is designed to allow a person to monitor his or her voice from the receiver to determine how loudly to speak. Unfortunately the sidetone also picks up background noise that is present in the listener's local environment and routes it though the receiver causing added interference to the speech signal. Holmes, Frank, and Stoker (1983) found the telephone listening ability of normal-hearing subjects was increased significantly in background noise when the sidetone was eliminated by either occluding the transmitter with the palm of the subject's hand or by electronically disengaging the transmitter, as can be accomplished with the mute button available on some telephones. The noise background caused by the sidetone can also be detrimental for listeners with hearing impairments. Occluding the microphone or using the mute feature in noisy environments can be particularly beneficial for these individuals.

DEVICE OPTIONS FOR TELEPHONE USE

Despite the limitations of the telephone system, options are available to help a person with hearing impairment improve his or her telephone-listening abilities. These include use of various types of telephone amplifiers and of hearing aids and other assistive listening devices (ALDs) coupled either acoustically or electromagnetically to a telephone.

Replacement Amplifier Handsets

One of the most effective means of amplifying the telephone signal is through the use of telephone amplifier replacement handsets. These relatively inexpensive handsets may be used with modular-type telephones with detachable receivers. Several models of these devices are available (figure 2). The most common types provide up to 20 dB of amplification, although some models can provide as much as 40 dB of gain. Some styles use a rotary volume adjustment, whereas others use a fingertip touch-control switch that automatically returns to normal volume when the handset is hung up. This feature is preferable for telephones used by several different people. It prevents a person with normal hearing from having to lower the volume when using the handset after a person with a hearing impairment. Other models have push-button or slide switches offering discrete volume choices.

Use of these handsets has been found to improve word reception abilities of listeners with hearing impairment significantly (Holmes

Figure 2.   Replacement amplifier handsets for modular-type telephones.

1985; Holmes and Frank 1984). Seventy-three percent of SHHH members surveyed stated they used amplifiers on their telephones and 45% of them found the amplifiers met their telephone-listening needs (Kepler, Terry, and Sweetman 1992). Pichora-Fuller (1981) found that all but one of 61 patients who tried an amplifier handset reported some benefit from its use.

In-Line Telephone Amplifiers

Another type of amplifier can be placed in-line on some modular-type telephones (Slager 1989). These amplifiers can be powered by a telephone line or by such an external power source as transformers or batteries. They are installed in a system between the base of the telephone and the handset cord (figure 3). These tend to be less convenient than

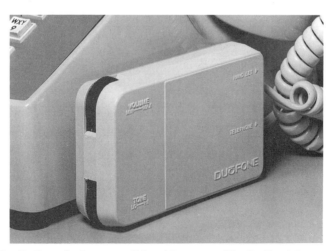

Figure 3.   In-line telephone amplifier. (Courtesy of Radio Shack.)

the replacement handsets, however they are usually less expensive. A potential source of difficulty with in-line amplifiers is that reception of the user's voice may be reduced to the person on the other line by those amplifiers powered by a telephone line (Compton 1991).

Portable Amplifiers

Battery-powered portable amplifiers are designed to be coupled to a telephone either acoustically or magnetically. Those that are coupled magnetically must be used with hearing aid compatible telephones. These devices are used by simply strapping the amplifier to the handset receiver (figure 4).

Telephones with Built-in Amplifiers

Telephones can also be purchased with built-in amplifiers and other special features for a listener with a hearing loss. Some units provide amplification similar to that of the replacement handsets, whereas others are designed to amplify the signal with a high-frequency emphasis. Such additional features as a low-frequency ringer, visual ringer light, and mute buttons are also standard on some models (figure 5). These telephones are now available for both residents and businesses.

Many public telephones are now equipped with built-in amplifiers (figure 6). As of January 26, 1992, the Americans with Disabilities Act (1990) mandates that twenty-five percent of public telephones in newly constructed buildings and facilities must have amplifiers for individuals with hearing loss [4.1.3(17)(b)]. In accordance with the ADA, employers must acquire or modify equipment, within reason, for individuals with hearing loss. Telephone amplifiers are considered to be a reasonable accommodation for hearing-impaired employees (Hablutzel, Hablutzel, and Gillio 1992).

Figure 4.   Battery-powered portable amplifier. (Courtesy of Radio Shack.)

Figure 5.   Telephone with built-in amplifier, visual ringer light, and mute button. (Courtesy of Williams Sound Corp.)

Figure 6.   Symbols for public telephones with amplifiers.

Hearing Aids and Telephones

Hearing aids may be used for telephone listening by employing either acoustic or magnetic coupling (figure 7). The term *acoustic coupling* implies that a telephone receiver is held to the microphone of a hearing aid, which picks up an acoustic signal from the telephone. The signal is then passed through the usual hearing aid circuitry and directed to the ear. Acoustic coupling may create a feedback problem, especially with high-gain instruments, because the receiver physically obstructs leakage of sound from the ear canal and deflects it to the hearing aid microphone. This feedback often can be reduced or eliminated by using a foam telepad attached to the receiver. This increases the distance between telephone and hearing aid, thus reducing the chance of reamplification of the signal. An additional problem with acoustic coupling occurs when the microphone of a hearing aid picks up background noise from the environment and from the telephone sidetone.

Hearing aids equipped with a telecoil or T-switch can be coupled to the telephone magnetically. A telecoil system takes advantage of the nonuniform radiation of electromagnetic energy from a telephone receiver (Gladstone 1975). It then converts this energy into electric energy, amplifies the signal, and transduces the signal back into an acoustic signal that is directed to the ear. Magnetic coupling avoids the problem of acoustic background noise being picked up from the hearing aid microphone, but it still amplifies any noise originating from the sidetone.

Not all telephones are compatible with hearing telephones. In the 1970s the L-type telephone receiver was designed for greater cost efficiency and longer life. Unfortunately, these receivers had a reduced amount of magnetic leakage, causing incompatibility between the L-type receiver and hearing aid telecoils (Gladstone 1975). The Hearing

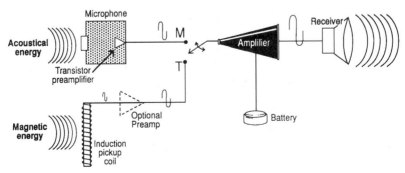

Figure 7. Simplified block diagram of a hearing aid with M-T switch for magnetic or acoustic coupling. (Courtesy of D. Preves, Argosy Electronics.)

Aid Compatibility Act of 1988 (Public Law 100-394) mandated that all new corded telephones manufactured or imported for use in the United States be hearing aid compatible after July 16, 1989, and that all cordless telephones be compatible after August 16, 1991 (Compton 1991).

In order to achieve maximum benefit from magnetic coupling, a telephone handset must be held on or close to a hearing aid (figure 8). Orientation of the telecoil in the hearing aid itself may require that an individual holds the handset at an odd angle to achieve maximum amplification. Induction coils that are oriented either horizontally or at a diagonal are optimal for telephone use. Those that are oriented vertically are best for audio room loops and neck loops but not for telephone use (figure 9) (Preves 1992).

Although much attention has been given to the use of magnetic coupling of hearing aids and telephones, its benefits have not been demonstrated in the literature or in clinical practice. Pichora-Fuller (1981) found that only twelve percent of her patients surveyed received benefit from hearing aid use on the telephone. In 1990, May, Upfold, and Battaglia reported on a survey of 244 hearing aid users. Their in-the-ear, in-the-canal, and behind-the-ear hearing aid users all indicated poor performance on the telephone using their hearing aids

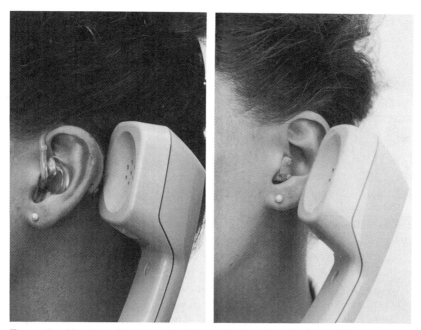

Figure 8.   Hearing aids magnetically coupled to the telephone. (Courtesy of Phonak, Inc.)

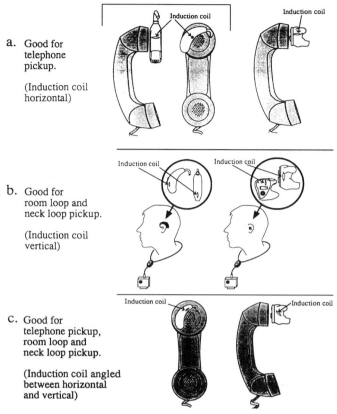

a. Good for telephone pickup.

(Induction coil horizontal)

b. Good for room loop and neck loop pickup.

(Induction coil vertical)

c. Good for telephone pickup, room loop and neck loop pickup.

(Induction coil angled between horizontal and vertical)

Figure 9. Optimal position for the hearing aid telecoil. (Courtesy of D. Preves, Argosy Electronics.)

with or without magnetic coupling. Likewise, Kepler, Terry, and Sweetman (1992) found that of the 55% of the SHHH members surveyed who used their hearing aids coupled with a telephone, 70% found it to be problematic. Poor performance of hearing aid use with telephones could be caused by a number of factors, including quality of the telecoil, magnetic field of the telephone, or position of the handset in relation to the hearing aid's telecoil. Telecoil frequency response curves for most hearing aids are poorer than the microphone response (Rodriguez, Meyers, and Holmes 1991).

Tannahill (1983) evaluated two different hearing aids and two different telephone receivers with normal-hearing listeners. The word recognition abilities of the listeners were significantly decreased with the use of telecoil coupling. Holmes (1985) found no significant differences in the telephone listening abilities of 19 subjects with mild-to-moderately severe, sensorineural hearing losses when the subjects were unaided or aided with either acoustic or magnetic coupling to the telephone. In another investigation 45 subjects were evaluated using unaided and aided (acoustic coupling) conditions (Holmes and Frank 1984). No significant differences were shown between conditions. In addition, use of electromagnetic adapters with a telephone has been investigated using magnetic coupling (Holmes and Chase 1985). Subjects had significantly poorer word recognition scores when using a telecoil alone coupled to a telephone, than when using an unaided telephone or a telecoil with an adapter.

One reason for the poor performance of hearing aids magnetically coupled to telephones may stem from lack of information available to the hearing aid dispenser on the electroacoustic characteristics of a hearing aid telecoil. Studies have demonstrated that switching a hearing aid from microphone to telecoil negatively affects the hearing aid's electroacoustic performance characteristics (Rodriguez, Holmes, and Gerhardt 1985; Rodriguez, Meyers, and Holmes 1991). Moreover differences between microphone and telecoil characteristics vary depending on the hearing aid manufacturer and the model. Figure 10 shows telecoil and microphone real-ear reponses of two different behind-the-ear hearing aids worn by the same subject. Note that one hearing aid's telecoil response is fairly similar to the microphone response, whereas the other is very different. To date, the standard measurement for telecoils (ANSI S3.22, 1987) uses only one frequency (1000 Hz). No information on the frequency response of telecoils is provided routinely by manufacturers. Currently, the ANSI S3-43 working group is investigating new standards for hearing telecoil measurements that would be more comprehensive (see Lederman, Chapter 2, this volume).

Another problem with the use of hearing aids coupled either acoustically or magnetically to the telephone is the lack of information available on interaction of the frequency response of a telephone with the frequency response of a hearing aid. Optimal hearing aid gain for telephone use may be very different from the optimal gain for standard face-to-face communication. Rodriguez et al. (in press) recently evaluated real-ear aided response typically preferred by a group of hearing-impaired listeners using acoustic and magnetic coupling conditions. They found that the subjects preferred a more gradually rising or flat frequency response during telephone listening conditions than that recommended by the most common gain formulas used for face-to-face communication. Therefore, the filtering effect of a telephone receiver altered the hearing aid response preferred by listeners. Programmable hearing aids are now available that offer listeners several frequency responses for different listening situations. It may be advantageous for dispensers to set one option with a flat frequency to be used specifically for telephone listening.

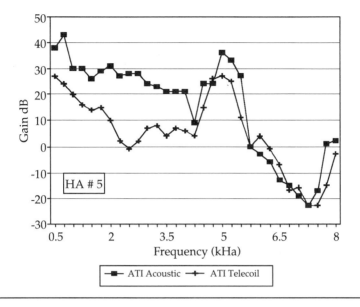

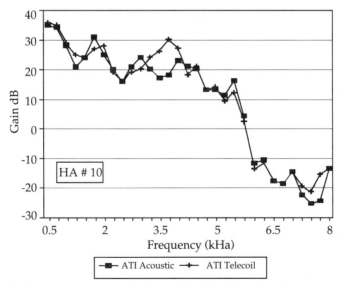

Figure 10.   Representative sample of one subject's real-ear insertion response (REIR) values of two different behind-the-ear hearing aids under acoustic versus telecoil coupling conditions. The top panel shows a hearing aid with the typically observed poorer performance under electromagnetic coupling conditions. The bottom panel illustrates REIR curves for a hearing aid that was able to achieve similar characteristics under acoustic and electromagnetic coupling conditions. (Reprinted with permission from: *The Volta Review*, Volume 93, No. 2, "Hearing Aid Performance Under Acoustic and Electromagnetic Coupling Conditions" p. 92. Copyright 1991, by the Alexander Graham Bell Association for the Deaf, 3417 Volta Place, NW, Washington, DC 20007.)

Personal Assistive Listening Devices and Telephones

Telephone interfaces are available to link personal hardwired and wireless assistive listening devices (ALDs) to the telephone (figure 11). These can be used with earphones or in conjunction with a person's hearing aids (Compton 1991). Many ALDs can also provide needed amplification of a telephone signal by the user's simply holding the handset upside down with its microphone at the mouth and placing the ALD's microphone at the receiver of the telephone.

COUNSELING NEEDS

Although telephone use is an important part of today's society, it is often overlooked in the auditory rehabilitative process of persons with hearing loss. Many individuals are fitted with hearing aids equipped with telecoils and are never instructed as to their function (Centa 1992). Professionals must become more informed on telephone options available to a hearing-impaired consumer. This information then must be passed on to their patients. McCarthy, Culpepper, and Winstead (1983) demonstrated lack of awareness about telephone aids in a survey of 50 hearing-impaired adults. Only 43% were familiar with telephone amplifiers.

Protocols must be developed to assist practitioners in evaluating the telephone needs of their hearing-impaired consumers. Appropriate assistive devices must be made available, and counseling on their

Figure 11.   Personal assistive listening device coupled to the telephone using an interface. (Courtesy of Williams Sound Corp.)

use with and without hearing aids is necessary. Technologies to assist hearing-impaired listeners on the telephone offer many options, but their benefits cannot be realized without education of both hearing-health-care professionals and hearing-impaired people.

REFERENCES

American National Standards Institute. 1987. *Specifications of Hearing Aid Characteristics* (ANSI S3.22). New York: American National Standards Institute.

Americans with Disabilities Act, P.L. 101-336. July 26, 1990 Title IV, Section 225, (d)(1)(c)-(g).

Centa, J. M. 1992. Telecoils: Federally mandated or voluntarily included? *Hearing Instruments* 43(8):43.

Compton, C. L. 1991. Clinical management of assistive technology users. In *The Vanderbilt Hearing Aid Report II*, eds. G. A. Studebaker, F. H. Bess, and L. B. Beck. Parkton, MD: York Press, Inc.

Erber, N. P. 1985. *Telephone Communication and Hearing Impairment.* San Diego, CA: College-Hill Press.

Fike, J. L., and Friend, G. E. 1983. *Understanding Telephone Electronics.* Dallas, TX: Texas Instruments Inc.

Gladstone, V. 1975. History and status of incompatability of hearing aids to telephones. *Journal of the American Speech and Hearing Association.* 17:103–104.

Hablutzel, N., Hablutzel, M. L., and Gillio, V. A. 1992. Legal overview. In *The Americans with Disabilities Act: Access and Accommodations—Guidelines for Human Resources, Rehabilitation, and Legal Professionals*, eds. N. Hablutzel and B. T. McMahon. Orlando, FL: Paul M. Deutsch Press.

Holmes, A. E. 1985. Acoustic vs. magnetic coupling for telephone listening of hearing-impaired subjects. *Volta Review* 87:215–22.

Holmes, A. E., and Chase, N. 1985. Listening ability with a telephone adapter. *Hearing Instruments* 36(9):16–17, 57.

Holmes, A. E., and Frank, T. 1984. Telephone listening ability for hearing-impaired individuals. *Ear and Hearing* 5(2):96–100.

Holmes, A. E., Frank, T., and Stoker, R. 1983. Telephone listening ability in a noisy background. *Ear and Hearing* 4:88–90.

Ingles, A., and Tuffnell, W. 1951. An improved telephone set. *Bell Systems Technical Journal* 30:239–70.

Kepler, L. J., Terry, M., and Sweetman, R. H. 1992. Telephone usage in the hearing-impaired population. *Ear and Hearing* 13:311–19.

May, A. E., Upfold, L. J., and Battaglia, J. A. 1990. The advantages and disadvantages of ITC, ITE and BTE hearing aids: Diary and interview reports from elderly users. *British Journal of Audiology* 24:301–309.

McCarthy, P., Culpepper, N. B., and Winstead, T. G. 1983. Hearing-Impaired consumers' awareness and attitudes regarding auditory assistive devices. Paper presented at annual convention of American Speech-Language-Hearing Association, November, Cincinnati, OH.

Pichora-Fuller, M. K. 1981. Use of telephone amplifying devices by the hearing-impaired. *Journal of Otolaryngology* 10:210–18.

Preves, D. A. 1992. Status of induction pickup in hearing aids for telephone and loop application. Paper presented at the 4th International Congress of Hard of Hearing People, August, Minneapolis, MN.

Richards, D. L. 1973. *Telecommunication by Speech*. London: Newnes-Butterworths.

Rodriguez, G., Holmes, A. E., and Gerhardt, K. 1985. Microphone vs. telecoil performance characteristics. *Hearing Instruments* 36(9):22–24, 57.

Rodriguez, G., Holmes, A. E., DiSarno, N. J., and Kaplan, H. (In press). Preferred hearing aid response characteristics under acoustic and telecoil coupling conditions.

Rodriguez, G., Meyers, C., and Holmes, A. E. 1991. Hearing aid performance under acoustic and electromagnetic coupling conditions. *Volta Review* 93:89–95.

Slager, R. D. 1989. Romancing the phone: The adventure continues. *Seminars in Hearing* 10(1):42-56.

Stoker, R. 1982. Telecommunications technology and the hearing impaired: Recent research trends and a look into the future. *Volta Review* 84:147–55.

Tannahill, J. C. 1983. Performance characteristics for hearing aid microphone versus telephone and telephone/telecoil reception modes. *Journal of Speech and Hearing Research* 26:195–201.

Visual Transformations

Chapter • 10

Sign and Oral Interpreters
The Who, What, When, and How

Janet L. Bailey

One way to ensure effective communication for deaf and hard of hearing people is to provide qualified interpreters. The Americans with Disabilities Act (ADA) lists interpreters under the generic heading of "Auxiliary Aids and Services." The Department of Justice has defined auxiliary aids as *"Qualified Interpreters,* note-takers, transcription services, written materials, telephone handset amplifiers, assistive listening devices, assistive listening systems, open and closed captioning, telecommunications devices for deaf persons (TDD), video-text displays, or other effective methods of making aurally delivered materials available to individuals with hearing impairments" (p.III-78).

As an employer, a government service, or a public entity, you will be asked to identify not only the need but the appropriate and most effective solution to the problem of communication with deaf and hard of hearing people. Before you can be expected to know *when* or *how*, it is imperative that we begin with a basic understanding of interpreting.

HISTORICAL PERSPECTIVE

An interpreter acts as a communication link between deaf and hearing persons who want to interact. An interpreter conveys messages using the mode of communication most readily understood by the persons involved. This can be spoken English, sign language (ASL, Manually Coded English), cued speech, or speechreading.

Interpreting is a relatively new field. Its historical roots are found in the charitable work of family members, friends, and teachers who "helped" when necessary. These "helpers" were not educated as interpreters and were often not paid for their efforts. As educational opportunities increased, job markets opened up, legislation supported equal access, and deaf and hard of hearing people began to require interpreters in more areas. It was clear that volunteer helpers could not keep up with the expanding needs or the rising expectations of consumers who were becoming more sophisticated.

In 1964, the Federal government awarded a grant to Ball State Teachers College in Muncie, Indiana to conduct a workshop on "Interpreting for the Deaf." This workshop was attended by approximately 75 persons, most of whom were educators or deaf leaders of the day. In a June 6, 1964 letter from William McClure, Workshop Director, participants were told that workshop objectives included, "...establishing standards for interpreters which could lead to a registry of *qualified* (emphasis mine) interpreters." Even in 1964 it seems that Mr. McClure and his colleagues recognized a "great and continuing demand for the services of interpreters."

During this historic meeting, the National Registry of Interpreters for the Deaf (RID) was formed to help educate the public about the interpreting process and to establish a registry of all interpreters nationwide. The RID worked closely with the National Association of the Deaf (NAD) to build trust between deaf consumers and professional interpreters. Toward that end they established and promulgated a code of ethics that set forth a standard for ethical behavior that professional interpreters were bound to follow. RID's Code of Ethics (see Appendix A) is now looked to as the model for the field.

Consumers using the Registry needed more than a list of names. They wanted a means to identify the skills of those listed. In response, RID members worked to adopt standards of excellence by establishing evaluation and certification programs that are recognized nationwide (see Appendix B).

Certification offers consumers an indication of the individual interpreter's competence. The holder of a national certificate has successfully completed a testing system, has been evaluated by specially trained professional interpreters and consumers, and has demonstrated a minimum level of competency.

As the only certifying body for sign and oral interpreters since 1972, RID has maintained an aggressive stance in monitoring test effectiveness. To assure the highest degree of validity and reliability, RID revised its testing instruments and launched a new and improved evaluation in 1988. In 1992, three RID task forces began work on developing standards and blueprints for examinations that would lead to new certifications for legal, deaf, and oral interpreters.

RID's years of experience and extensive work in the area of certification and evaluation is second to none. Although various government agencies, businesses, professional organizations, and consumer groups have developed screening tools and evaluation measurements, the RID's system is the only one supported by a formal National Grievance structure (recently renamed the Ethical Practices Board) designed to protect consumers and regulate its own members. Furthermore, to assure that RID certified interpreters keep abreast of new developments in the field, RID members have mandated a Certification Maintenance Program that will require all certified members to participate in continuing education activities.

RECOGNIZING THE NEED

Deaf and hard of hearing persons may be able to communicate effectively in some situations by speechreading, written notes, or through an assistive listening device. However, this type of communication can be time consuming, cumbersome, and frustrating for all involved. When these methods prove to be difficult or impossible, it may be best to consider bringing in a professional interpreter. When in doubt as to whether or not an interpreter is necessary, it is a good idea to consult with the deaf or hard of hearing consumer.

The ADA suggests that public accommodations rely on disabled individuals whenever possible in determining the best way to provide equal access. Knowing your customers and clients is not only a good way to remain in compliance with the law—it is, moreover, good simple business practice. Getting acquainted with the deaf community in your area is an excellent way to tap into a new market.

Once you have determined that you should provide interpreter services, you must decide what type of services are needed. It is important to ask the following questions.

When . . . do we need an interpreter?
Who . . . or what kind of interpreter do we need?
How . . . do we find a qualified interpreter?
What . . . should we know about working with an interpreter?

WHEN . . . do we need an interpreter?

In some instances, the situation will dictate the need for an interpreter. For example, if you run a book store, you probably will not need to provide an interpreter for a deaf individual who is an occasional drop-in customer, browsing or selecting a book. If, however, you present an author's lecture and book-signing reception, a deaf consumer might request that you provide an interpreter.

Speechreading is not always effective, and not all deaf and hard of hearing persons are skilled at it. If you own a toy store and a hard of hearing consumer asks you where the dolls are, you can be sure that normal speech, gesturing, and pointing will suffice. If, on the other hand, you are an attorney negotiating a contract with a deaf businessman, there will be few environmental clues to assist in the communication. Contract negotiations are apt to be much more complex, providing too many opportunities for misunderstandings that could result in serious damage.

Common sense can be used as a guide in many instances. Consider the individuals involved, the subject matter, the environment, and the complexity of the situation. Again, you are urged to consult with individual consumers and representatives from disability groups.

WHO . . . or what kind of interpreter do we need?

You will need to know what type of interpreter is needed. Will your consumers need the services of a sign language (ASL or Manually Coded English), oral, tactile, or cued speech interpreter?

Sign Language Interpreters. These interpreters facilitate communication between nonsigners and consumers who use American Sign Language or one of the manual codes for English. The term *interpreter* is used when interpreting between American Sign Language and spoken English. A "transliterator" provides the message in one of several manually coded forms of English, which borrows some signs from ASL but presents them in English word order.

Oral Interpreters. These interpreters serve deaf and hard of hearing persons who rely on speechreading for receptive communication. They may be able to communicate effectively in most one-to-one situations, but they may require an interpreter when speechreading is difficult or impossible. Oral interpreters mouth the spoken message silently, repeating the words along with the speaker. Often this is done verbatim, but an interpreter may choose to paraphrase if the words are difficult to discern from the lips. Group discussions where dialog switches quickly from one speaker to another, auditorium presentations, or theatrical events may be easier to comprehend with the aid of an oral interpreter. If a consumer prefers, an interpreter will also speechread them and voice for hearing consumers.

Certain physical characteristics of a speaker can pose problems for clear communication. A beard or a mustache, for example, can make speechreading impossible.

Tactile Interpreters.   These interpreters sign or fingerspell into the hands of deaf-blind consumers. This normally requires at least one interpreter for each consumer, and a team if the interpreting assignment lasts for more than an hour (see Chapter 13, this volume).

Cued Speech Transliterators.   These cued speech transliterators combine the use of cued speech with a well-defined transliterator role, providing access to the language and culture of the mainstream. Fleetwood and Metzger (1990) explain in their book, *Cued Speech Transliteration* that, "Cued Speech is a visual representation of the pieces of a spoken language, called phonemes." Spoken phonemes are easily distinguished by hearing people because each has a different sound. However, these sounds look similar or identical when they are speechread. Cued speech uses handshapes and hand placements in combination with the natural mouth movements of speech to provide visible markers that allow a consumer to see the distinctions that hearing people hear.

What is the difference between an interpreter or a transliterator?
The term *interpreter* refers to the act of translating between two languages; in this case ASL and English. As Frishburg (1990) explains, we are talking about "interpreting between people who do not share the same language or mode of communication" (p. 1). The term *transliterator*, on the other hand, identifies an individual who facilitates communication between a spoken language and a signed or cued form of the same language; in this case English. In this chapter the term *interpreter* is used to refer to both interpreters and transliterators.

If you are providing an interpreter for an individual or a specific group of consumers, you can always ask them for assistance in providing appropriate service. They might also be able to help by suggesting possible resources.

Offering general interpreting services, however, can be more tricky. For example, if you have a theatre or concert hall and you want to expand your audience, you might opt to provide interpreters for certain nights of each performance. Many theatres now offer sign language interpreters for one or two performances and audio loops for hard of hearing nonsigners who might attend. The key to successful provision of general audience coverage is to know your potential community. Meet with interested theatregoers and with local interpreting agencies.

HOW . . . do we find a qualified interpreter?

Within the past ten years there has been a dramatic shift in *how* to retain an interpreter. It used to be that most practitioners worked as freelance interpreters or as sole proprietors. Those wishing to hire an interpreter were expected to locate a listing of individual interpreters, contact them directly (often by leaving many messages on their answering machines), and negotiate individual contracts for each assignment. This meant that the responsibility of guaranteeing appropriate services was left with the individual or entity requesting the service.

In recent years, consumers have been able to provide services by calling on interpreter service providers (ISP) who have centralized the administrative and scheduling duties previously handled by individuals. These firms might be nonprofit social service agencies, state offices, or private businesses.

The advent of the ISP has allowed interpreting to be conducted in a more professional manner. The ISP has assumed administrative responsibilities that individual interpreters were unable to attend to because they were busy interpreting. By contracting with a competent and respected ISP, you can be sure that you will be in compliance with existing legislation. Most ISPs follow national standards for interpreter competence, and they should be able to verify the qualifications and credentials of their personnel.

Interpreters may maintain their freelance status and contract with an agency for assignments, or they may be employed as staff interpreters. Whatever the arrangement is between the interpreter and the ISP, this change in procurement procedures has meant that interpreters are free to focus on their craft, while the ISP attends to marketing, scheduling, and accounting.

Although the ISPs have helped streamline procurement of interpreters in many areas, they have not solved the problem of supply and demand that has affected our profession in recent years. It is a common practice in many of our busiest cities to require two weeks notice to locate a qualified interpreter. You will want to locate a service provider in your area and find out how they arrange for interpreters before you have a specific need. Make your requests with plenty of notice whenever possible. Many ISPs now have programs that respond to emergency needs. This, however, is normally reserved for true emergencies.

Since passage of the ADA, the demand for interpreting services has far exceeded available resources. The RID estimates that there are approximately 2,000 certified interpreters throughout the United States and another 1,000 associate members working toward certification.

With over 75 interpreter preparation programs (IPPs) throughout the nation, Mary Wright, President of the Conference of Interpreter Trainers (CIT)[1] estimates that there are approximately 1,000 students

---

[1] Conference of Interpreter Trainers, P.O. Box 257, 3262 Superior Lane, Bowie, MD 20715

graduating each year. The influx of these new graduates alone cannot solve the shortage that we currently face. Many graduates find work in mainstream school settings and do not make themselves available to meet the needs of the general community.

Interpreter preparation programs offer experiential practicums and internships to assist their students. Mentor programs have been established by businesses and professional associations to allow veteran interpreters to assist novice interpreters in their orientation to professional life. More programs aimed at developing resources and recruiting new talent to the field are needed.

WHAT . . . should we know about working with an interpreter?

Requesting an Interpreter. When calling to hire an interpreter you should be prepared to provide the following information.

1. Date, time, and length of assignment
2. Location of the event
3. On-site contact person and phone number
4. Description of the event (i.e., technical training, informal staff meeting, dental appointment, performing arts, contract negotiations, legal deposition)
5. Additional information to assist in appropriate placement (i.e., black tie event, the interpreter will be videotaped and included in an insert, or special security arrangements will have to be made ahead of time)
6. Name(s) and interpreting preference of the consumer(s), when possible
7. Complete billing information

How Many Interpreters Will We Need? Because of recent findings by the Occupational Safety and Health Review Commission related to repetitive motion injuries and because these syndromes have become a growing problem for interpreters of the deaf, interpreter service providers are taking steps to reduce the occupational stress that may affect these workers. Policies have been adjusted regarding the number of interpreters required for specific assignments. The following guidelines should be followed to assure occupational safety for all working interpreters.

I. One interpreter is usually sufficient for:
    A. Most meetings of two hours or less
    B. Situations that do not require continuous interpreting and offer adequate rest periods for the interpreter
II. Two or more interpreters (a team) are needed when:

A. A platform assignment extends for more than one hour
B. A meeting/training/conference extends for more than two hours
C. A meeting of any length requires more than one method of communication, (e.g., sign language, oral interpreting, cued speech, or tactile interpreting)
D. The deaf consumers are situated in two or more physical locations (e.g., a speaker on a dais and audience members)

Where Should the Interpreter(s) Be Placed? Interpreters are often asked by the contact where they should be positioned to assure comfort for all involved. The interpreter may offer suggestions, but it is always advisable to involve the consumers in the discussion as well. As experienced consumers of interpreting services, deaf people are in the best position to explain their personal needs and preferences.

The interpreter's position will depend on the type of assignment. In general, you should consider where the deaf consumers will be seated and whether they will have a clear sight line to both the interpreters and the main speakers. Keep in mind that watching an interpreter can cause eye strain and fatigue for deaf consumers. Be sure the interpreter is not standing in front of a bright window or a glittering crystal chandelier.

For small group or one-to-one situations (such as physician visits, counseling sessions, bank transactions, or social conversations) the interpreter will most likely move to the side of the hearing consumer so that the deaf or hard of hearing person will have the most direct sight lines to both the interpreter and the speaker.

For auditorium seating, it is normally best to reserve seats in the front rows for deaf and hard of hearing people. If you are not sure where deaf consumers will be seated, you should probably make certain that the interpreter will be positioned on stage with a light. If you know that the consumers will be sitting on the right side of the audience, you might be able to place the interpreter on the floor in front of the stage so that the consumers will be able to see the speaker just above the interpreter.

For formal panel presentations, where there may be deaf audience members as well as those offering testimony, it is best to offer at least two teams of interpreters. For government hearings where members of Congress sit on a raised platform facing a table placed in front of the public gallery, I have found it best to have one team focused on those presenting testimony and another off to the side for the audience. When deaf persons are seated in front of the panel you may need two sign-to-voice interpreters seated directly in front of the panel facing the speakers. Microphones normally placed on this table will need to be stretched to reach the voice interpreters. (These logistical details will be dealt with prior to the actual hearing or presentation.) If you have only one voice-to-sign team they should move directly behind the panel during testimony so that any questions will be caught by the deaf presenter. By having an interpreter directly behind the seated panel member, the deaf presenter will also have an automatic cue as to who is asking the questions.

Once again, we urge you to work with deaf and hard of hearing consumers and professional interpreters when trying to arrange for the most appropriate positioning for specific situations.

Pre-assignment Preparations. Interpreter preparation is always important. Work with your interpreters to assure a smooth communication process.

1. Offer agenda, lists of topics, and specialized vocabularies. Send these prior to the date of the assignment, when possible. A list of pertinent names, acronyms, and jargon is helpful, even if received just before a meeting.
2. Consider working out special arrangements when you plan to use overhead projectors, films, or demonstrations. Will the lights need to be dimmed or turned off? If so, what provisions can you make to assure sufficient lighting for the interpreter?
3. It is extremely helpful to provide printed copies of scripts, lyrics, or poems when performances will be part of a presentation.

Tips for Working with your Interpreter.

I. You should look directly at the deaf or hard of hearing person, although the interpreter is speaking to you. The deaf consumer will probably look back and forth from the speaker to the interpreter to get both content and effect of the message.

II. Speak directly to the deaf person using the first person. Do not say, "Tell him . . ." or "Ask her to . . .".

III. The interpreter should be positioned as close to the hearing speaker as possible so that the deaf consumer can see both easily.

IV. The interpreter will not be maintaining eye contact with the speaker and will need to be able to hear clearly what is being said. Encourage hearing group members to speak clearly and audibly.

    A. Speakers often use such directions as, "Look at the last paragraph on page 32," or "As you see here in the graph . . .". Deaf and hard of hearing participants cannot focus on a speaker and look at a page simultaneously. They will have to choose between the printed display and the interpreter.

    B. Speakers often refer to a blackboard or flip chart and offer such visual cues as, "Look at that . . ." or "Just like the one on the right." These phrases are next to impossible to interpret, and they may cause the interpreter to interrupt your presentation to ask for clarification.

V.   Because interpreters must translate from one language to another, there will be a time lag in their delivery.

   A.   Responses to questions and laughter at the end of a joke may be delayed.

   B.   Deaf and hard of hearing members of a group discussion may have trouble interjecting ideas or concerns. If the group agrees that members will raise their hands to be recognized by a chair or leader it will make the group discussion more accessible to all. It will also reduce the chaos often experienced in free-for-all discussions, where the loudest and most forceful voice wins the floor.

Remember, if the interpreter is having trouble following the discusion, it is likely that hearing members are having the same problem.

VI.   Speed is always a problem. Most meetings start with introductions. The names of participants often are given rapidly, leaving the interpreter struggling to catch up, and the deaf individual unsure of who is attending the meeting. When speakers read a passage, they tend to present material much faster than if they were speaking extemporaneously.

VII.   Repeat questions or comments from the audience when necessary.

VIII. Be aware that some things may be lost in translation.

   A.   Puns, sound-based words or phrases, or jokes that depend on the understanding of a sound may be lost.

   B.   Phrases that are strongly based in hearing culture may be lost (e.g., "Sock it to me" or "Get a life!")

   C.   Interfering noises, mumbling, or accents always create problems for interpreters.

IX.   Technical language, references to personal names, and acronyms can be problems for the interpreter.

X.   Understand that the interpreter may have to interrupt to clarify or rephrase.

XI.   Interpreters are trained professionals who are bound by a code of ethical behavior that incorporates such important principles as confidentiality and neutrality (see RID code of ethics, Appendix A). It is important that interpreters remain in their interpreting role. The following will assist them in this.

   A.   Do not have private conversations with the interpreter during the process. Do not say things to the interpreter that you do not want "interpreted."

B. Do not ask the interpreter's opinion.
C. Do not ask the interpreter to become involved in the event, the discussion, or the procedure.

Your interpreters will probably arrive early to prepare for the assignment. They will want to meet with consumers, the contact person, and any scheduled presenters to help everyone understand the process. Depending on the situation, they will offer some of the suggestions listed above.

SUMMARY

If you determine that you need an interpreter, consult with deaf and hard of hearing consumers to identify *when* and *what kind* of services will be needed. Consult with deaf consumers on *where* to locate qualified interpreters. Use common sense and courtesy when you think about *what* you should do when working with an interpreter.

And finally, you should offer your consumers a way of evaluating your services. The evaluation will be a valuable test of effectiveness for both the interpreting and your program or service.

REFERENCES

Americans with Disabilities Act Handbook (Title III-§36.303 (b) (1) Auxiliary aids and services. U.S. Printing Office, Washington, DC: EEDC and the U.S. Department of Justice (EEOC-BK-19).

Fleetwood, E., and Metzger, M. 1990. *Cued Speech Transliteration: Theory and Application*. Silver Spring, MD: Calliope Press. (Original quotation appeared on p. 19. Revised in letter by authors Feb. 17 & 20, 1992.)

Frishberg, N. 1990. *Interpreting: An Introduction (Rev. Ed.)*. Silver Spring, MD: Registry of Interpreters for the Deaf, Inc.

APPENDIX A: CODE OF ETHICS OF THE REGISTRY OF INTERPRETERS FOR THE DEAF, INC.

Introduction

The Registry of Interpreters for the Deaf, Inc., has set forth the following principles of ethical behavior to protect and guide interpreters and transliterators and hearing and deaf consumers. Underlying these principles is the desire to ensure for all the right to communicate. This Code of Ethics applies to all members of the Registry of Interpreters for the Deaf, Inc., and to all certified non-members.

Interpreter/transliterator shall keep all assignment-related information strictly confidential.

Interpreter/transliterator shall render the message faithfully, always conveying the content and spirit of the speaker, using language most readily understood by the person(s) whom they serve.

Interpreter/transliterator shall not counsel, advise, or interject personal opinions.

Interpreter/transliterator shall accept assignments using discretion with regard to skill, setting, and the consumers involved.

Interpreter/transliterator shall request compensation for services in a professional and judicious manner.

Interpreter/transliterator shall function in a manner appropriate to the situation.

Interpreter/transliterator shall strive to further knowledge and skills through participation in workshops, professional meetings, interaction with professional colleagues, and reading of current literature in the field.

Interpreter/transliterator, by virtue of membership in or certification by the RID, Inc., shall strive to maintain high professional standards in compliance with the code of ethics.

APPENDIX B: CERTIFICATION CLASSIFICATIONS

The following is a listing of the national certificates recognized by the Registry of Interpreters for the Deaf, Inc. (RID). This list is offered with the express permission of the National RID:

CI, Certificate of Interpretation—ability to interpret between American Sign Language and spoken English in both sign-to-voice and voice-to-sign.

CT, Certificate of Transliteration—ability to transliterate between signed English and spoken English in both sign-to-voice and voice-to-sign.

CSC, Comprehensive Skills Certificate—ability to interpret between American Sign Language and English and to transliterate between English and a signed code for English.

MCSC, Master Comprehensive Skills Certificate—advanced ability to interpret between American Sign Language and English and to transliterate between English and a signed code for English.

RSC, Reverse Skills Certificate—ability to interpret between American Sign Language and signed English or transliterate between English and a signed code for English. The interpreter is deaf or hard of hearing, and interpretation/transliteration is rendered in American Sign Language, spoken English, a signed code for English or written English.

TC, Transliteration Certificate—ability to transliterate between English and a signed code for English.

IC, Interpretation Certificate—ability to interpret between American Sign Language and English.

SC:L, Specialist Certificate: Legal—indicates specialized knowledge of legal settings and greater familiarity with language used in the legal system.

PROV:SC:L, Provisional Specialist Certificate: Legal—indicates that the individual has taken specialized training in legal interpreting.

SC:PA, Specialist Certificate: Performing Arts—indicates specialized knowledge and greater familiarity of performing arts settings and a demonstrated ability to successfully convey artistic materials.

OIC:C, Oral Interpreter Certificate: Comprehensive—ability to paraphrase and transliterate a spoken message from a hearing person to a deaf or hard of hearing person and ability to understand and repeat the message and content of the speech and mouth movements of the deaf or hard of hearing person.

OIC:S/V, Oral Interpreter Certificate: Spoken to Visible—ability to paraphrase and transliterate a spoken message from a hearing person to a deaf or hard of hearing person.

OIC:V/S, Oral Interpreter Certificate: Visible to Spoken—ability to understand the speech and silent mouth movements of a deaf or hard of hearing person and to repeat the message for a hearing person.

The Cued Speech Training, Evaluation, Certification Unit (TECUnit) also offers testing and national certification. Currently, individuals who pass the Cued Speech Transliterator National Certification Examination (CSTNCE) secure a certificate at one of three levels. Eventually, only the top level certificate will be available. That certificate is defined as:

TSC, Transliteration Skills Certificate—ability to transliterate/transphone, between spoken English, phonemically similar spoken languages, and Cued Speech from both voice-to-cue and cue-to-voice without influencing the communicative participants, the content and context of their messages, or the situation, its progress, and its outcome.

APPENDIX C: RID GUIDE TO INTERPRETER SERVICE ORGANIZATIONS

RID maintains an updated state by state listing of Interpreter Service Providers. Contact:

Registry of Interpreters for the Deaf, Inc.
8719 Colesville Road, Suite 310
Silver Spring, MD 20910
(301) 608-0050

Chapter • 11

Developments in Real-time Speech-to-Text Communication for People with Impaired Hearing

E. Ross Stuckless

This chapter offers a practical overview of real-time speech-to-text communication from the perspective of its present and projected applications for people with impaired hearing and for others with whom they communicate, including both hearing and hearing-impaired people.

For good reason, many people with impaired hearing impatiently await the availablity of a portable device—figuratively a small black box—which will automatically and instantly convert spoken language into type for the hearing-impaired person to read. When it materializes, such a device will give people with impaired hearing much greater access to society's prevalent mode of communication—spoken language. Deaf people in particular will be able to directly and independently "tune in" on spoken communication in school, in the community, and in the workplace to a degree not hitherto possible.

It should be noted at the outset that this chapter discusses the specific sequence of converting speech into text. It mentions only briefly its reciprocal, the conversion of text into speech, commonly known as "speech synthesis." From both scientific/technologic and linguistic perspectives, there is little resemblance between automatic speech recognition and speech synthesis. As a man-and-machine activity, speech-to-text is a much more formidable task than text-to-speech. For deaf people who do not have intelligible speech, speech synthesis (text-to-speech) probably has numerous useful applications. However, to date it has not gained much attention.

DEFINITION OF REAL-TIME SPEECH-TO-TEXT

For the purposes of this chapter, real-time speech-to-text is defined as *the transcription of words that make up spoken language accurately into text momentarily after their utterance.*

Speech-to-Text

There is nothing new in speech-to-text communication per se. Written English, as most other languages, is derived from its spoken form. Arguably, most of us draw on what has variously been called implicit, subvocal, covert, or internal speech (Conrad 1979) to aid us in composing and organizing much of what we write or type (Williams 1987).

With the introduction of the audiotape recorder it has become a simple, although time-consuming, task to transcribe speech into handwritten or typed text. Moreover, with the appropriate equipment and skills, it is a relatively straightforward task to create conventional captions for film or video. But both of these situations require that we first record the spoken activity, giving us time to transcribe it onto paper or some other medium.

Real Time

There is no common standard for "real time" as used here. The concept has at least two elements. One has to do with the "live" quality of the communication activity itself, as in face-to-face interactions, speeches and lectures, and live television broadcasts of programs such as *Monday Night Football* and *The MacNeil/Lehrer News Hour*. Relatively little of the 400 or more hours of captioning produced for broadcast television in a given week is done in real time because most programs and their captions can be, and are, prerecorded, at least those on the national networks.

A second element in the use of the term "real time" pertains to the delay between the spoken utterance of a word or a brief passage and its appearance as text. Some readers may recall an early effort to reduce the delay in adding captions to "live" events, in this case reporting the news on television. In 1973, WGBH in Boston received permission from ABC News to videotape its nightly news, produce captions, and replay the captioned version later the same night over the Public Broadcasting System nationwide. Although this did not constitute real time, it was aired within about five hours after the initial broadcast, and it represented a notable achievement at the time. Today we measure delay in seconds.

The definition of real-time speech-to-text, as presented in this chapter, speaks of text being generated "momentarily after" the spoken utterance. Under real-time conditions, text cannot appear in absolute synchronism with the speech it has "copied." This is analogous to the conversion of speech into signs by a sign interpreter. There must be some delay for several reasons, one of which has to do with maintaining accuracy of text.

For most communication involving real-time speech-to-text, the briefer the interval between the spoken utterance and its appearance as text, the more beneficial for a reader of the text, particularly in one-to-one and group situations involving interaction. For example, deaf people who use an interpreter in a group meeting or classroom discussion are penalized as active participants when they receive information several seconds after their hearing peers. This problem can and does also occur when reading real-time speech-to-text in a similar situation.

For practical purposes, the delay between a spoken utterance; for example, a word or a phrase, and its display in print should be under three seconds. However, it should be pointed out that there may be a trade off between brevity of delay and accuracy. This point is elaborated later in the chapter.

Accuracy

Accuracy is based on the similarity between language as spoken and its textual counterpart. Completely accurate text is free of: (1) word errors; and (2) word deletions. Word additions are a third potential source of error, but they occur so infrequently that they are not considered here. A method for assessing accuracy and classifying errors is described later in the chapter.

Word Errors. When we write or type a message of any importance, we normally have the opportunity to proofread the text for errors and to make appropriate corrections before the message is shared with others. Under real-time conditions, any error in transcription from speech into text is likely to be read before it can be corrected. Depending on the nature and frequency of errors, this can be quite distracting, reducing or even negating the value of the text to a reader (Stuckless 1982a).

Word Deletions. To qualify fully as real-time speech-to-text, a transcription from speech into text should be word-for-word as spoken. A reader of the text should be able to read the language of a speaker fully, without deletions. But this will depend on whether the text can be generated at the same speed as the speed of the spoken language from which it is derived. That cannot occur unless the speed with which text is generated can match the speed of the spoken language from which it is derived.

At present, virtually all systems for transcription of speech to text require a manual interface of some kind. Either the speaker or a "recorder" must manually record speech as it is being uttered. Today's manual interfaces include variations on handwriting, use of a keyboard, or use of a stenographic machine. Although limited in function, today's automatic speech recognition systems are indeed automatic, requiring little or no manual recording.

Speed

To understand better the problem of word deletion in generating complete text from spoken language in real time, we must examine the speed of spoken language, as compared to our ability to record it manually or have it recorded automatically.

The speed of *speech* varies considerably from person to person and from one circumstance to another. Turn (1974) has reported between 2.0 and 3.6 words per second (120 to 215 words per minute [wpm]) as average rates of spontaneous speech. I investigated the speaking rate of ten college professors as they lectured. The average speaking rate of the group, inclusive of brief pauses beween utterances, was approximately 150 wpm. Actual rates varied from 112 wpm for one instructor to 180 for another. A third instructor, while speaking at an average rate of 143 wpm, was observed to peak at the rate of 260 wpm over a brief 35-word utterance (Stuckless 1983).

However, *handwriting* is much slower than speech. The reader is invited to check his or her own cursive handwriting rate by copying as much of a newspaper editorial as he or she can within one minute, while keeping the manuscript reasonably legible for another reader. Then count the number of words recorded. I did this and averaged 32 words over three trials. At less than 25% of the speed of speech, under most conditions handwriting cannot offer a complete and accurate transcription of speech in real time. However, we can take abbreviated notes.

The speed of *typing* on a conventional typewriter, a computer keyboard, or a telecommunication device for the deaf (TDD) can be considerably faster than handwriting. With training and practice, many of us can reach 60 to 80 wpm or more with reasonable accuracy, approximating 50% of the speed of speech. This offers a major improvement over handwriting, but it remains unsatisfactory for producing a complete and satisfactory transcription of speech into text in real time. A secondary consideration is how long a particular rate can be sustained by the typist (or the handwriter) before fatigue begins to reduce speed and accuracy.

A third type of manual interface is the *stenographic machine*, a device commonly used by court reporters. When coupled with a skilled stenotypist and a computer with appropriate software, this is by far the most efficient manual interface available today for transcribing speech into text in real time. Well-qualified stenotypists can record speech for an hour or more at rates well over 150 wpm, and with relatively few errors.

APPLICATIONS FOR PEOPLE WITH IMPAIRED HEARING

Applications of real-time speech-to-text for people with impaired hearing have been reported for more than 100 years. In 1882, a machine designed to transcribe speech automatically was described in the *American Journal of Otology* (Blake 1882). Called a *glossograph*, it involved placing a false palate in the speaker's mouth, with levers extending outside the mouth to transcribe the speaker's spoken language in readable form. At least, that was the idea. A year later, this device was discussed by the editor of the *American Annals of the Deaf* (Fay 1883), who made the following comment:

> We fear that the glossograph, ingenious as it is, is too inconvenient in its *modus operandi* to be of much real service; but in view of this and the other wonderful inventions of recent years, it is not unreasonable to hope that some instrument will yet be contrived that will record human utterance automatically without inconvenience or annoyance to the speaker. (pp. 67–68)

When Fay wrote about this instrument, he almost certainly envisioned a device independently capable of reading spoken language and converting it to text—automatic, real-time speech-to-text. More than 100 years later, this dream remains elusive, at least as an automatic activity.

HANDWRITING

Handwriting is a common application of real-time speech-to-text communication used by many people who are deaf, particularly for face-to-face interaction with people who hear. It is a substitute for the use of speech and hearing, or of signs, to communicate. Unfortunately, its value has gone largely unrecognized except by its immediate users.

Face-to-face Interactive Communication

Simply described, one person writes a note on a pad or sheet of paper and passes it to the second person for a response, exchanging thoughts as in other forms of face-to-face communication. The purpose of the communication may be as straightforward as seeking directions or asking for an item while shopping, or as complex as a job interview or a parent/teacher conference.

Unlike some other applications of real-time speech-to-text, in this instance the speaker records his or her own text manually. Unfortunately, the slow rate of communication imposed by the speed of handwriting (approximately 30 wpm) reduces the usefulness of what otherwise seems to be a relatively efficient process. In order to offset this limitation partially, the messages exchanged are usually brief and often without full sentence structure.

Based on their demographic studies of the deaf population in the metropolitan Washington, DC area and later throughout the United States, Schein (1968), and Schein and Delk (1974) reported that 50% or more of all deaf people surveyed wrote only, or wrote in combination with speech, in their interactions with sales clerks and with their supervisors on the job. Supervisors tended to respond in a similar manner.

More recently, Foster (1992) conducted in-depth interviews with 20 supervisors of deaf employees in a variety of work settings. Consistent with the findings of Schein and Delk, she reported that 11 of the 20 indicated the use of reading and writing, either in a supportive or primary role, to communicate with their deaf employees. Several said their deaf employees routinely carried a pad and pencil with them in order to facilitate communication. Foster identified two positive features of face-to-face handwritten communication for deaf people and their supervisors: (1) they generally are understood correctly; and (2) if speech, lipreading, or sign language are not feasible, "everyone expects to write and the awkwardness of trying to decide what will work is avoided" (p. 69). But Foster noted that some reservations were expressed also, even among those supervisors who generally supported the use of writing to communicate with their deaf employees. First was the inconvenience and the time required in comparison with that of speaking and listening. Second, the comment was made that some people, both hearing and deaf, are not comfortable using reading and writing for face-to-face communication—among these people, some are simply poor readers and poor writers.

At this time, for real-time face-to-face communication between two nonsigning people, handwriting, although technically primitive, more closely approximates speech than any other system yet devised, because of its combination of the following attributes.

1. Interactive, expressive/receptive capabilities
2. Immediate availability
3. Absolute portability
4. No special skill needed beyond ability to read and write in a shared language; e.g., English, Spanish
5. No third party needed for transcription purposes
6. No dollar cost.

Its major disadvantage remains its inefficiency in terms of speed.

Real-time Handwritten Notes in Meetings

Up to this point, our discussion of handwriting has focused on its "conversational," one-to-one applications. It has not touched on group meetings and larger gatherings wherein a person with impaired hearing is present. As a substitute for a manual or oral interpreter, someone in a meeting may sit beside this person and write running text that the hearing-impaired person can scan. An overhead projector is sometimes used for the same purpose, particularly when several people with hearing impairments are present.

Unlike a "conversational" situation in which speech rate is reduced to writing rate, discussion in a meeting conducted at normal speaking rates cannot be recorded fully in writing in real time. In effect, a person with impaired hearing does not have full access to a meeting—only to notes. As a practical matter, these notes do not allow a person to participate actively in a meeting. Woodcock (1992), an executive who became deaf in her teens, has this to say about the subject.

> Pencil and paper function quite well between two people, but can't do anything useful for one deafened person in a group. With handwritten notes, the deafened person will know the general topics that the others have been talking about, approximately two topics after they have moved along. Access is not provided by making sure the deafened person knows "they are talking about the budget now." Access is knowing exactly what is being said about the budget. Access is being able to join in the conversation and contribute. (pp. 50)

This is not to imply that notes or full minutes of a meeting are without worth to its participants who are hearing-impaired, but that their value during (not after) the meeting is likely to be minimal. Special amplification, interpreters, and reasonable courtesies on the part of other group members are likely to be much more effective relative to the full participation of the person with impaired hearing. Obviously he or she should be consulted on this matter.

TYPING

Face-to-face Interactive Communication

Although typing can double or even triple the speed of handwriting, it is not often used for face-to-face interactive communication. A keyboard and its attendant text display cannot compete with pencil and paper for portability, convenience, or cost. An exception to this may be its application to instruction in a classroom, where portability is not a major consideration.

As a pilot study, in the 1970s I placed two Digalog keyboard terminals that were wired to TV screens in each of several classrooms in a school for the deaf. The teachers and their students were encouraged to use these instruments to communicate during a portion of each day for several weeks, and the teachers maintained logs of their activities and observations. Teachers and students were generally motivated to use the system.

Since then, more elaborate computer-based face-to-face interactive communication systems have been developed for classroom use. At least one of these, Electronic Networks for Interaction (ENFI)[1], has been used extensively in classes of students with hearing impairments. Classroom-based systems that encourage deaf students to engage in dialog with the teacher and each other through text in real time warrant more attention than they have received to date, particularly for the purpose of English language development.

Distance Interactive Communication

We cannot overstate the value of the Telecommunications Device for the Deaf (TDD), or the more recent designation, Text Telephone (TT), as a substitute for the conventional or amplified telephone among hundreds of thousands of people with severe-to-profound hearing impairments. The TT is mentioned briefly in this chapter as an application of real-time speech-to-text. For more detailed information, the reader is referred to Castle (Chapter 8, this volume).

The TT is a portable device with a keyboard and an electronic visual text display for sending and receiving messages, and an acoustic coupler to connect the device to a telephone line. Some TTs also incorporate a paper feed, providing a printed record of both outgoing and incoming communications. Like a conventional telephone, the TT can be used to make local, national, and international calls. Its major limitation has been that a person being called must also have

---

[1]For more information about ENFI, contact Joy Peyton, Center for Applied Linguistics, 1118 22nd St., NW, Washington, DC 20037.

direct access to a TT, thereby precluding such day-to-day applications as making appointments, calling the office to report in sick, and in many areas of the country, dialing 911.

Recently this problem has been alleviated considerably with the acquisition of TTs by more public and private offices and services. There is a growing awareness nationally of hearing-impaired people as citizens and taxpayers, and as consumers of goods and services. No doubt the 1992 Americans with Disabilities Act (ADA) is also beginning to have its intended effect on communication access.

Creation of relay services is another major development in distance interactive communication. Based on the realization that people with and without access to TTs needed to interact with one another, relay services began to emerge around the United States in the 1970s and early 1980s. Procedurally, two people—one with access to a TT, another with standard telephone access—can communicate with one another through a relay operator who relays their messages back and forth using both a TT and voice. This is a practical example of the joint application of speech-to-text and text-to-speech concepts, both in real time.

Where these relays are well established, they have demonstrated their worth. Many have expanded and merged into statewide systems. Capping these systems, and in compliance with provisions in the ADA, a nationwide system was put in place in 1993.

COMPUTER-ASSISTED NOTE TAKING

It should be pointed out that note taking, as we use the term here, is just that—the recording of notes, not verbatim text. Unlike handwritten note taking, computer-assisted note taking typically involves the use of a keyboard. The most apparent differences between the two are speed and legibility, because of the efficiency of typing over handwriting[2].

There are differences in the sophistication of various computer-assisted note taking systems that distinguish one from another in what they can do. As might be expected, the more advanced systems generally require higher skill levels on the part of the note taker, or more accurately, the note taker/typist/computer operator.

A major benefit of all computer-based note taking systems is that the real-time text can be stored in the computer and retrieved later for printing—with or without editing—and distribution to participants in the meeting or class.

---

[2]A new "pen-based" computer technology is now emerging, in which the keyboard is replaced by a writing instrument; e.g., "Go-Point," Microsoft's "Pen Windows," Apple's "Newton." Several features of this technology may have implications relative to our particular interest in communication.

Basic Systems

By way of hardware and software, basic systems require a low-cost computer/keyboard, word processing software, for example, Word-Perfect (making sure in advance that the note taker is familiar with the particular software), and a monitor. If the number of people watching the monitor makes viewing difficult, either multiple monitors (or a larger monitor) can be used, or a data projection panel with an overhead projector and screen. It should be noted that the room lighting must usually be dimmed when using a projector, creating problems in speechreading or following an interpreter unless special lighting is provided.

The most essential requisites for the note taker are that he or she be a good typist and be familiar with the particular word processing software to be used. I recommend an article by Virvan (1991) on the use of basic systems by hard of hearing people.[3]

A variation on this basic system, HI-LINC, allows the typist to view and alter what he or she has typed before sending it to the reader (Kyle 1990).[4] Another variation, InstaCap,[5] permits recalling prerecorded text and other such macros as peoples' names alongside real-time text. As a secondary benefit, the InstaCap package also includes a caption generator for display purposes.

Woodcock (1992) considers such basic systems as just described to be useful as a supplement to listening and speechreading for hard of hearing and oral deaf people, and as a way of retrieving notes after a meeting (assuming the notes are printed out and distributed following the meeting). However, she considers these systems inadequate as substitutes for interpreters or *verbatim* real-time speech-to-text in providing what she calls "true access" to meetings.

This having been said, I am reminded of a deaf colleague who returned from an out-of-state trip with praise for having seen virtually error-free verbatim real-time speech-to-text being created by a typist on a basic system. Curious, I contacted the person responsible, expecting to learn about some remarkable new software development. Instead I was told the software was a basic commercial package called

---

[3]A two-page description of a basic system assembled by Self Help for Hard of Hearing People, Inc. is also recommended. This description includes equipment specifications and instructions to the note taker, and is available by writing to SHHH, 7800 Wisconsin Ave., Bethesda, MD 20814. Ask for information on "SHHH captioned note taking."

[4]HI-LINC is an abbreviation of Hearing-impaired Live Information through Computers. For information about HI-LINC, contact the Centre for Deaf Studies, University of Bristol, 22 Berkeley Square, Bristol BS8 1HP, England, UK.

[5]Inquiries about InstaCap should be sent to RW Thompson Systems, Cavanmore Rd., Carp, Ontario K0A 1L0, Canada.

"Word Processing for Kids." But when I then asked about the qualifications of the typist, I was told that the typist was able to maintain an almost error-free rate of 110 wpm on a standard computer keyboard over an extended period. This rate apparently sufficed for verbatim recording of that particular meeting.

This anecdote highlights the importance of expertise on the part of a typist, note taker, or stenographer in transcribing speech into text in real time, technology notwithstanding.

Systems Using Abbreviations

Several systems have incorporated abbreviations of words and phrases into their computers. This is done to reduce the number of keystrokes needed to produce text, enabling a typist to approximate more closely the speed of speech and come closer to producing complete text in real time.

Probably representing the first effort in this direction for communication with people with hearing impairments, a project was initiated in 1972 under the sponsorship of the Royal National Institute for the Deaf in England to produce verbatim text from speech in real time. This included reconfiguration of a computer keyboard and abbreviation of words with two-letter codes, requiring considerable special training on the part of the typist (G. Hales, personal communication, June 18, 1973). Although the actual outcomes were disappointing, it was a beginning.

Today, a number of software packages are available that permit entering abbreviations on the keyboard to recall full words or phrases on the screen: for example, "exp" for "experience." Most provide a list of frequently used words and their abbreviations, together with instructions for adding words and their abbreviations to the list. Several of these packages are listed below.

MindReader
    Brown Bag Software
    2155 S. Bascom Ave.
    Campbell, CA 95008
Gregg Computer Shorthand
    Glencoe
    936 Eastwind Dr.
    Westerville, OH 43081
Shorthand Plus
    Computer Resource & Support
    100 North 80 East
    Provo, UT 84606

Speedwriting Conversion Software
    Spencer Data Processing
    934 Edgemoor Rd.
    Cherry Hill, NJ 08034
Quickey
    Trace Research & Development Center
    University of Wisconsin
    1500 Highland Ave.
    Madison, WI 537105
Productivity Plus
    Productivity Software International, Inc.
    1220 Broadway
    New York, NY 10001

A system using abbreviations currently is under development at the National Technical Institute for the Deaf (NTID) in Rochester, NY. Called "C-Print," its goal is to transcribe spoken language at up to 150 wpm, with accuracy at 95% or better (Henderson, McKee, and Stinson 1990; Wilson 1992). This system uses Productivity Plus software to enable the typist to enter an abbreviation and produce a full word from a special NTID-designed dictionary within the computer. WordPerfect software is used to process the text.

Based on recognized phonetic principles, a set of 40 rules was developed at NTID to govern how abbreviations are formed for the dictionary. With training, the typist should be able to apply these rules to select a word from, and to add new words to, the C-Print dictionary (as an alternative to the formidable task of attempting to memorize a list of several thousand abbreviations for instant recall). Training manuals were developed based on these 40 rules.

After 105 hours of training on this system, a group of six paid trainees whose conventional typing speeds varied between 20 and 60 wpm, substantially increased their word production rates using the system. One typist who continued to work with the system during field testing with college classes reached a word production speed of around 120 wpm, approximately double her conventional typing rate.

Developers believe that in order for a typist to reach 150 wpm and 95% accuracy with C-Print, he or she should be a skilled conventional typist with typing speeds of 70 to 80 wpm, and have a good "ear" for phonetics/speech sounds before beginning training on the system. They also estimate the need for about ten weeks of training at six hours per day (J. Henderson, personal communication, Oct. 28, 1992). More work remains to be done on this system before it is determined whether it can, in fact, be used in practical situations to record speech in real time as *verbatim text*. Failing this, a less desirable but more feasible option might be its consideration and evaluation as an advanced computer-assisted *note taking* system. As with all computer-based systems requiring a typist or stenographer for word entry, we need to determine how long this person can sustain speed and accuracy comfortably, without relief, under normal working conditions.

As we see, considerable training and practice time is required on the part of the typist before he or she becomes skilled in the use of computer "abbreviation" software. This in turn has implications for the provider and user of the service in terms of the length and cost of training, and the availability of suitably qualified typists. If these prove to be major problems, using a steno/computer system as described in the following section may help. It should be added, however, that cost and availability problems are often associated with staffing for steno/computer systems also.

There may be another option: redesign of the keyboard. The QWERTY keyboard as designed for the typewriter more than 100 years ago and subsequently incorporated into the computer keyboard, has severe design deficiencies, both ergonomically and for speed, for today's computer applications (Feder 1992; Gopher and Raij 1988). It is conceivable that a redesigned keyboard, used in tandem with a system using abbreviations, could add further to speed of transcription and help the typist avoid discomfort and fatigue.

Several computer-compatible keyboard redesigns are presently available, including the BAT. The BAT has only seven keys which are pressed simultaneously in different combinations to produce letters and commands.[6] This particular device resembles the stenographic machine in the sense that they both use a "chording" principle that entails depressing several keys simultaneously. However, the resemblance ends there. Unlike the typist, the stenographer works essentially with sounds, not letters.

STENOGRAPHIC COMPUTER-ASSISTED SYSTEMS

Early History and Applications

Probably the earliest effort to link stenography with computers was made by IBM in the mid-1960s. The Central Intelligence Agency subsequently used an early system for rapid (not real-time) transcription of documents. In the late 1970s, the technology entered the public domain, initially to expedite the transcription of stenographic recordings into legal documents. Its first public application in real time may

---

[6]Information about the BAT is available through Infogrip, Inc., 5800 One Perkins Place, Suite 5-F, Baton Rouge, LA 70808.

have been its use in a New York State courtroom by an attorney who was hearing impaired.

In January 1982, virtually identical systems, produced by Translations Systems, Inc., came into use at the National Technical Institute for the Deaf (NTID) and at the National Captioning Institute (NCI), the former for research and evaluation with deaf students in regular college classes at Rochester Institute of Technology, and the latter for real-time closed captioning for television. In March of the same year, it was used in the Supreme Court by a deaf attorney. Its first application in a large meeting of hearing-impaired people occurred at the national convention of the Alexander Graham Bell Association in June 1984.

Over the ensuing decade, the major applications of these systems have continued to be: (1) real-time captioning for television at the network level; (2) real-time use with hearing-impaired students in regular classes with hearing peers, mostly at the college level; and (3) real-time coverage of meetings, usually of a national scope, attended by large numbers of hearing-impaired people. Among these three, the greatest growth has occured in real-time captioning for television, though seldom for locally produced programs.

The Reporter and the Stenographic Machine

In place of a keyboard, the stenographic computer-assisted system—the "steno system"—uses a stenographic machine (figure 1). Various versions of this device have been used for more than 100 years by court reporters and others with similar training to transcribe speech sounds phonetically into paper tape, in the form of letter codes (figure 2).

The steno machine has 24 keys. Most word codes, and some codes for common phrases, are formed by a single stroke, explaining in part the fact that highly skilled reporters can transcribe speech at rates well in excess of 200 wpm. While recording the voice of a person speaking at 150 wpm, the reporter is in fact pressing about 11 keys per second. Parenthetically, it might be noted that a single keying error can produce the kind of word "glitch" often seen by TV viewers of real-time closed captions (Miller 1989).

In ordinary writing and typing, we separate words with spaces. But in expressing ourselves through speech, we blend words into one another as in "BLENDINGONEWORDINTOANOTHER." A reporter, in a similar manner, records speech sounds without separating one word from another, contributing greatly to his or her recording speed.

At this point, conventional stenographic reporting and computer-assisted reporting go their separate ways. Without computer assistance, a reporter or someone else trained to read the tapes must then transcribe the tapes manually into commonly understood English.

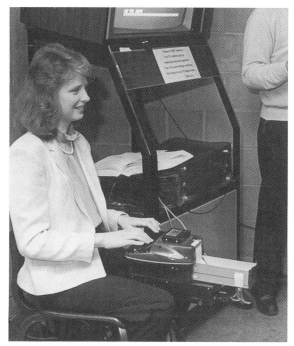

Figure 1.   Reporter and steno machine.

With computer assistance, this is done automatically, making real time transcription possible.

The Computer and Its Software

The fundamental function of the computer and its software is to convert the word codes sent to it by the reporter into accurate, correctly spelled text, and to put this text in a format suitable for being dis-

Words	Letter Codes
writing	WREUG
on	O
the	-T
machine's	PH-PB/APS
keyboard	KAOE/PWORD

Figure 2.   Words and letter codes.

played and read. A secondary task is to store the text so it can be retrieved and edited. In part, for this purpose, the computer has its own keyboard and monitor.

When the first of these systems became commercially available around 1980, it had most of the features present in newer systems. However, its hardware was basically nonportable and required frequent service, thus limiting its applications. Moreover, as might be expected of first generation technology, the equipment was quite costly. Today, however, most systems can operate using a standard laptop computer and software that costs around $5,000,[7] so this no longer represents a major cost in the overall system. Numerous systems are available, but some companies have adapted more than others for the real-time captioning market. In alphabetic order, Cheetah Systems (Fremont, CA), RapidText (Irvine, CA), and XScribe (San Diego, CA) are among the leaders.

Essentially the software consists of three parts: (1) a computer dictionary; (2) a program that selects words from the dictionary based on a particular logic and set of rules; and (3) a word processing program that organizes these words in a particular format and performs other basic editing tasks (figure 3).

At present, there is no uniform theory or standard for how reporters should code English words (figure 2). Numerous existing theories are being reviewed by the National Court Reporters Association (NCRA) to promote greater consistency among reporters. Also, over time each reporter develops his or her own shortcuts in writing codes for certain words and phrases.

For these reasons, every reporter who uses a computer for reporting, develops his or her own computerized personal dictionary that contains a vocabulary of words paired with their letter codes. Over time, some reporters build personal dictionaries of 200,000 words or more. If one reporter were to use another reporter's personal dictionary, the transcription error rate would likely be unacceptable.

A comment should be made about the importance of preparation time on the part of the reporter for entering new vocabulary into his or her dictionary. Let us assume the reporter is asked to cover a lecture or a speech on a topic with which he or she is not familiar. In this case, it is probable that some of the vocabulary (e.g., peoples' names, technical terms) is not already in his or her dictionary. If so, the probability of error in writing these words is high. The alternative is to avoid writing that word, producing an error of deletion (Stuckless and Matter 1982).

---

[7]Not included are costs of such equipment as the steno machine, printer, and visual display. Costs of the latter can vary greatly, from the price of a standard TV set to $10,000 or more for high-quality large-screen projection equipment.

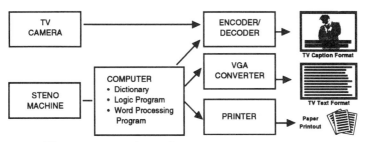

Figure 3.   The computer and its software.

This is why reporters often ask for a list of names and technical words likely to be used in a particular presentation or discussion. With adequate preparation time, these can be added to the reporter's dictionary in advance of their use by the speaker.

The second major software component is the logic program. This program is responsible for the selection of the correct words from the dictionary. Earlier we discussed the fact that a reporter does not indicate the boundaries of a given word, using the following example: "BLENDINGONEWORDINTOANOTHER." If the computer called up each group of letters as they first formed a word, we would see the following, which makes no sense: BLEND IN GO NEW OR DIN TO AN OTHER. Because the actual codes are phonetic in nature, the words called up can be even stranger, sometimes amusing, and occasionally embarrassing.

To avoid this kind of error, the logic program uses an algorithm that causes the computer to delay making a decision on a word until several phonetic codes have been entered by the reporter. This gives the computer more context to work with. Although this can result in a two or three second delay between the moment a word is spoken and when it appears on the display, this is preferable to a briefer delay that produces a considerably lower level of accuracy (Stuckless 1982b).

Homonyms can be common sources of errors. The logic program may include a limited ability to parse a sentence and identify the more likely of two words as in "It was *too* far *to* walk." The program may also include some additional elementary functions related to grammar analyses.

The computer's third major software component is its *word processing program*. This consists of a standard software package such as WordPerfect, and is used for formating and editing text in preparation for its printing as hard copy.

The Display

Depending on the purpose, text can be displayed in real time as text alone or as captions superimposed on a television picture. For a number of practical reasons, *text format* is the format of choice in classrooms and most other "live" meetings in which people with hearing impairments participate.

The text format is a considerably less complex and expensive display than the caption format. If the system's computer has a video output, it can be connected directly to a conventional television set for text display. If not, a computer/video interface is needed to link the computer and the television set.

A typical TV text format provides up to 16 lines of text, and up to 40 characters per line. For readability, a "scroll over" arrangement for introducing new lines of text is preferred over a "pop up" arrangement in which text moves up the screen (Stuckless 1978).

A *caption format* is used primarily for closed-caption broadcast television and involves superimposing captions electronically on a TV picture.

The TV caption format is sometimes used in large meetings of, or including, people with hearing impairments. In some respects this situation resembles a regular TV production, calling for a TV camera, other production equipment, and at least one technician. Special lighting may also be required. However, unlike conventional TV, the display in this instance is usually in the form of large screen projection (figure 4). The major advantage of this type of display is that people in the audience who are hearing impaired can read the text of what the speaker is saying at a considerable distance. Also, they can see the speaker "close up" for speechreading purposes.

When linked to a printer, stenographic computer-assisted systems can also print and display text as *hard copy*. Though not a real-time application, it can be particularly useful in an educational setting where the hard copy can be shared with students for review and study purposes. This has been demonstrated among hearing-impaired college students (Stinson et al. 1988). Regarding this application of the system for instruction, I recommend that a reporter edit the text for errors, punctuation, and paragraphing before sharing the hard copy with students.

Reporters covering noninstructional meetings for people with hearing impairments are often asked by an officer or another member of the group for a hard copy of the meeting. This is not a good practice unless the reporter and all the participants agree to it beforehand.

Choosing the Equipment

Reporters usually refer to their systems as "computer-aided transcription" (CAT) systems. A number of different systems have been developed for the stenographic reporting profession and most, if not all, of these can be used more or less as described in this chapter. It should be noted, however, that each system operates differently and requires a

Figure 4.   Peter Jennings of ABC World News Tonight presenting a real-time captioned speech to an audience of deaf students at NTID.

period of special training on the part of a reporter. It is equally important to know that the dictionary a reporter develops for one system does not automatically work on another—although for a fee, most CAT vendors can transfer a personal dictionary from one system to another.

A reporter's preference for a particular system generally depends on the system he or she has trained on and/or is accustomed to working with. Most, if not all, reporters and agencies offering real-time services for hearing-impaired people have and use their own equipment, including software, when reporting. Additional arrangements may be needed for acquiring and installing an appropriate display.

Unless an organization expects to make considerable use of a system and to employ or retain a reporter on a steady basis, it probably should not purchase its own system. However, if it chooses to acquire its own system, it should give consideration to the availability of local reporters experienced with a particular system and/or what would be involved in bringing an otherwise well-qualified reporter "up to speed" on a chosen system.

Several particular systems are preferred over others by real-time captioners and their agencies because of special features incorporated into these systems and the services offered by their manufacturers. Before an organization purchases a system for providing real-time speech-to-text services to hearing-impaired people locally, it is advisable first to ask experienced real-time captioners and their agencies what systems they are using and why.

Seeking a Reporter

The reporter remains by far the most important part of the system for its services to people with hearing impairments. There are approximately 40,000 active stenographic reporters in the United States, of whom perhaps one-third are employed as court reporters. Most are members of the NCRA.[8]

In a given year, an estimated 12,000 students enroll in 100 NCRA-approved reporter-training programs nationwide, typically for two years or more of intensive training. Several levels of certification as a Registered Professional Reporter (RPR) are available for those who qualify, based primarily on reporting speed and accuracy, and extending from 225 wpm up to 260 wpm. The NCRA recently established a certification program for reporters who transcribe from speech into text in real time, called "Certified Real-time Reporter."

A practical suggestion for those seeking to establish local real-time reporting services for hearing-impaired people is to focus the search for reporters on those who live within commuting distance of the needed service, at least in well-populated areas. It is recommended also that the search be limited to reporters with CAT experience and, depending on the circumstances, to those with their own equipment.

Local contacts and referrals might include hearing-impaired people and local organizations that have already used real-time reporting services for hearing-impaired people, reporters and reporting agencies in the area, and local and regional reporter-training programs. Also, the NCRA can be asked to provide a list of member reporters in the area who might be contacted.

As of early 1993 there were an estimated 90 full-time "real-time captioners" in the country. Many of these reporters are employed in four or five centers that specialize in producing real-time and prerecorded captions for national broadcast TV and entertainment videotape distribution. These people are sometimes available to cover meetings of hearing-impaired people, particularly at the national level, and the quality of their work generally is excellent; but unfortunately for most hearing-impaired individuals and groups at the local level, their availability is limited and the cost of their services is high. However, because of their contacts in the reporting community, these same people can be excellent referral sources.

Under the auspices of the NCRA, short courses are now being offered periodically thoughout the country to give already qualified reporters special training in providing real-time speech-to-text services

---

[8]The NCRA encourages its members to take an interest in real-time reporting for hearing-impaired people. Its address is 8224 Old Courthouse Road, Vienna, VA 22182-3808. Its major periodical is the *Journal of Court Reporting*.

npaired people. Special certification for real-time reporters
1 established. Interested reporters and prospective users of
orting services may contact the NCRA for more informa-
ese special training and certification programs.

her recent development, the Massachusetts Commission
' and Hard of Hearing, together with the Massachusetts
.eporters Association and a chapter of the Association of
ed Adults (ALDA), has established a full statewide ser-
ng approval for Massachusetts state agencies to pay for
-time reporting for hearing-impaired people in a manner
eir payment for interpreting services. Fee schedules for
agencies, as well as special training and certification procedures, have been developed within Massachusetts, making it the
first state to do so.

Choosing a Reporter

As indicated earlier, it is suggested that the candidate have experience
working with a CAT system and that he or she be able to provide his
or her own basic equipment. Given that such a candidate is available,
and available when needed, two basic questions should then be asked:
(1) is the candidate qualified; and (2) can the organization afford his or
her services?

There are many reporters nationwide with general reporting
skills but not at the level needed to qualify as court reporters. Some
leaders in the reporting field, and at least one company that develops
and markets CAT systems, believe that many of these more junior
reporters, and some reporting students, have the skills needed to meet
most real-time reporting needs for people with hearing impairments.
If this is so, it should substantially increase the pool of available
reporters and reduce the overall cost of real-time reporting for hearing-impaired people, particularly at the local level.

Others believe that real-time reporting for hearing-impaired children and adults, and particularly for those with difficulties in reading,
should be as error-free as possible, requiring a high skill level on the
part of the reporter. Research comparing the relative abilities of deaf
and hearing college students to identify and correct reporting errors
supports this latter position (Stuckless 1982a). Pending convincing evidence to the contrary, I support giving considerable weight to certification of the candidate at some level, although not necessarily at the
level of court reporter.

We should also discuss *cost* of the reporter. Well-qualified reporters can earn substantial incomes working in legal and business
communities, putting their normal reporting rates considerably above

the ability of most local public and private nonprofit organizations to pay. The fees paid to interpreters for providing communication services to deaf people provide a useful benchmark for reporters because their training and duties resemble one another in several ways. One difference is that, unlike interpreters, reporters must invest in several thousand dollars worth of equipment.

In connection with a particular project, I developed a suggested fee schedule in 1991 for part-time and occasional services for reporters covering local meetings and classes for hearing-impaired people. Except for reimbursement for equipment use, this schedule is similar to fees paid to freelance sign interpreters with advanced certification in New York City and in Rochester, New York. The use of similar payment schedules for real-time reporters and sign interpreters should encourage local organizations to choose between the two, based on hearing-impaired persons' needs and preferences rather than on costs.

Suggested Payment Schedule for Occasional and Part-time Real-time Reporting in Local Meetings and Classes of Hearing-impaired People

First two hours (including preparation and setup time)	$75
Each additional hour (including hourly ten minute break)	$35
Equipment charge, per hour (excluding display)	$15
Travel time and expenses	(negotiable)

The fees suggested here are considerably below those charged by most certified reporters; reporters and prospective users of their services obviously are free to negotiate for any fee acceptable to both. However, reporters' fees that are competitive with the fees of comparably qualified interpreters are likely to be more acceptable than those that are not. This has a persuasive logic, because the two services resemble each other so closely in their intent to convert speech, in real time, into a visual medium (i.e., text or signs) for ease of understanding by millions of people with hearing impairments.

Remote Systems

Before leaving this discussion of stenographic computer-assisted systems, we should say something about their applications if the reporter is at a distance from the actual communication. We live in an era when proliferating communication technologies are being coupled with each other increasingly, often with dramatic effects. Real-time stenographic reporting is no exception.

Since the introduction of real-time television captioning, reporters have been able to add live captions to a nationally televised event thousands of miles away. They watch and listen to the program in their own offices on their own television sets, and transmit their stenographic machine outputs via telephone line to a studio, where it is added to the picture in milliseconds. Linda Miller, a real-time reporter in Rochester, New York, sends real-time captions from her home, at 2 AM, Eastern Time, for a television program that originates and airs at 11 PM, Pacific Time, in Vancouver, British Columbia. She picks up the program via a satellite receiver on her property.

Dimitri Kanevsky and several associates at an IBM research center have been able to demonstrate linking hearing-impaired employees throughout the country with real-time reporters located in Chicago. A hearing-impaired employee participating in a local meeting calls the reporter, who in turn monitors the meeting and sends the real-time text to the employee over another telephone line. The employee then reads the text of the speech, within a second or two, on a TV or laptop computer screen. They have also demonstrated the system with cellular phones where fixed telephone lines are not available (Kanevsky et al. 1992). In many instances, major expansion of this kind of service might prove feasible and cost effective, particularly where local reporting services are not available.

AUTOMATIC SPEECH RECOGNITION

The various systems described up to this point require one of the following two conditions: (1) *a speaker* transcribes his or her own speech into written or typed text, adjusting speaking rate to match his or her rate of writing or typing; or (2) a speaker and a hearing-impaired reader utilize *a third party* to make the transcription in real time.

Early in this chapter we spoke of a third alternative, a figurative "small black box" that would *automatically* convert speech into text in real time. The reference, of course, was to automatic speech recognition (ASR). Actually, the term *automatic speech recognition* is more generic than the meaning we give to it in this chapter, and the applications more extensive (Lea 1980; Makhoul et al.1989; Waibel and Lee 1990). Although there is a degree of constancy at the recognition level, outputs vary considerably. For example, ASR can be used under structured conditions to control machines and to automate simple commercial transactions by telephone, demonstrating sophistication in its ability to analyze speech signals, but requiring little, if any, ability to process natural language.[9]

These kinds of applications receive considerable attention in commercial literature and public media, and some perform specific

[9]For more comprehensive conceptual treatments of ASR, see Lea (1980) and Waibel and Lee (1990), and for current information of a more technical nature, see the annual *Proceedings of the Speech and Natural Language Workshop*, sponsored by the Defence Advanced Research Projects Agency.

tasks well. However, their constraints are often overlooked or understated, leading to an erroneous expectation among many of us who are hearing impaired, and others of us who interact extensively with hearing-impaired people, that the "small black box" expected to print all the speech we encounter in our daily lives is just around the corner.

Indeed, while scientists make major progress in ASR, "grand challenges" in this direction remain before us as long-term goals (Makhoul et al. 1989). Some ASR researchers refer to our interest in automatic real-time speech-to-text as "automatic dictation/transcription." Of automatic dictation/transcription, Makhoul and his colleagues write, "The challenge lies in the system's ability to transcribe arbitrary spoken input with virtually unlimited vocabulary and types of sentence construction" (p. 464). Waibel and Lee (1990) make the same point when they say:

> Whereas these short-term applications will increase productivity and convenience, more evolved prototypes could in the long-run profoundly change our society. A futuristic application is the *dictation machine* that accurately transcribes arbitrary speech. Such a device can further be extended to an *automatic ear* that "hears" for the deaf. (p. 1)

Some believe that with appropriate resources, prototype ASR products can be ready for use by hearing-impaired people within five years (Hinton 1992). I am skeptical that enough will be known about the natural language processing side of ASR to make this possible within five years, but our difference of opinion may hinge on what we mean by "prototype." Later in the chapter I suggest several performance specifications for a fully functioned system for use by children and adults with hearing impairment.

The focus of this chapter is on speech-to-text, with hearing people serving as "speakers" and hearing-impaired people as "readers." At least one group of ASR researchers has tested deaf people as speakers to see if the varying qualities of deaf peoples' speech will permit them to access ASR functionally as speakers (Abdelhamied, Waldron, and Fox 1992). If so, this could lead to some useful outcomes for deaf people relative to interactive communication.

State of the Art

It is generally agreed that a model ASR system of the future will have three essential properties: (1) the ability to process continuous speech; (2) the capacity to recognize a large vocabulary; and (3) the ability to "understand" many speakers of the same language (i.e., speaker independence). We might add two performance conditions for each of the above: accuracy and real time.

Systems today can be differentiated on the basis of whether they recognize words in the context of naturally spoken language in which words blend together as continuous speech, or whether they recognize isolated words only. The latter, called *word recognition* systems, require brief pauses between each word, similar in some ways to the spaces between words in text. These systems can also be designed to recognize specific isolated phrases and brief sentences, and to treat them as if they were words.

Although this kind of system has the advantage of being much simpler to construct than a system designed to recognize continuous speech, it has the major disadvantage of forcing a speaker to speak in an artificial manner, generally at a rate of no more than 35 or 40 words per minute. Nevertheless, it can have practical applications with hearing-impaired people. For example, a word recognition system such as DragonDictate, now licensed to IBM as IBM VoiceType, may have several applications with children who are deaf, particularly to assist in their language and speech development.

Continuous speech recognition systems have not advanced at nearly the pace of word recognition systems, but their ultimate value to society and to hearing-impaired people in particular, promises to be much greater than the value of word recognition systems. Much of the early research on continuous speech recognition took words as the smallest unit of analysis, but we recognized that the virtually inexhaustible number of ways in which words can be combined in English mitigated against this approach, except for use with small, restricted vocabularies.

More recently, the phoneme (of which there are approximately 50 in English) has been generally adopted as a standard unit for continuous speech recognition, and statistical models have been developed around various combinations of phonemes.[10] One such model, the Hidden Markov Model (HMM), has had a major influence on the way scientists now view the recognition of continuous speech (Lee 1990). Today the prevailing view is that substantial progress is being made in the recognition of continuous speech.

The Oxford English Dictionary contains approximately 500,000 word entries, with an estimated 500,000 additional words, mostly technical and scientific, yet to be cataloged (McCrum, Cran, and MacNeil 1986). A *large vocabulary* ASR system may contain several thousand words, and the largest up to 100,000 words. The smallest may consist of only a few digits. *Small vocabulary* systems are inherently confining because the speaker must choose the words he or she uses from a limited selection.

---

[10]A particularly promising combination is the "triphone," which consists of a phoneme together with the two phonemes to its immediate left and right. Because of the number of possible combinations, triphones number in the thousands.

Unfortunately, large vocabulary systems present another kind of problem—the larger the pool of words in the computer's dictionary, the greater the likelihood of confusion and error. On the positive side, problems associated with storage and rapid retrieval of words from large vocabulary systems are easing with the development of faster, larger capacity computers.

Most ASR systems, and particularly those with large vocabularies, are *speaker-dependent*, that is, they are "trained" to recognize the voice of a particular speaker; if a second speaker wants to be understood, he or she must formally acquaint the system with his or her voice characteristics. Training may involve reading words aloud several times into the system from the list of words in the computer's dictionary. More recently it has become common to read a short list of phonetically balanced words or sentences from which the system makes generalizations about the speaker's voice characteristics.

As might be expected, speaker-dependent systems are generally more accurate than systems that are speaker-independent. However, speaker-dependent systems have the obvious disadvantage of being able to recognize only the speech of those who have taken the time to adjust the system to recognize their voices. This is a severe constraint for most practical purposes.[11]

As implied in the preceding paragraph, *speaker-independent* systems are more versatile than speaker-dependent systems. However, the price of versatility may be limited vocabulary and/or reduction in accuracy. The Hidden Markov Model (Lee 1990) has played a major role in recent developments in speaker independence, as was previously noted for continuous speech recognition, also. The most notable uses of speaker-independent systems have been in Voice Input/Voice Response telephone applications (Roe et al. 1991).

Today, some ASR systems are quite capable of processing *continuous speech*. Other systems can recognize *very large vocabularies*. Yet other systems demonstrate a high level of *speaker independence*. The challenge now is to incorporate all three of these capabilities into a single system. Raymond Kurzweil, a leading figure in word recognition, said:

> I expect it to be possible in the early 1990s to combine *any two* of these attributes in the same system. In other words, we will see large vocabulary systems that can handle continuous speech while still requiring training for each speaker; there will be speaker-independent systems that can handle continuous speech but only for small vocabularies; and so on. The Holy Grail of speech recognition is to combine *all three* of these abilities, as human speech recognition does. (Kurzweil 1990, p. 270)

---

[11]Speaker-dependent systems may have practical applications for hearing-impaired children in school and at home, where these children have frequent contact with a particular teacher or two, and with parents and siblings.

In this discussion of ASR, we have touched several times on *accuracy*. Our tolerance for error in the transcription of speech into text depends on the need for accuracy within a particular situation. When a physician describes to a hearing-impaired patient how frequently he or she should use a medication, or a teacher tells a student who is hearing-impaired what chapters in a textbook will be covered in an upcoming test, a much higher order of accuracy is required than in casual conversation.

We have said little about what constitutes real time with reference to ASR. We often read in the ASR literature that a given speech-to-text task required, as an example, 20 times real time, meaning that it took the system 20 times longer to analyze and process the speech input than it took the speaker to make the utterance. Obviously this time delay would be intolerable under most practical conditions.

Other variables also come into play in determining the usefulness of a particular ASR system for general use by hearing-impaired people. These include *portability*, *text legibility*, *exclusion of extraneous environmental noises*, and *cost*.

Desirable Performance Characteristics in an ASR System for People with Hearing Impairments

Using my applications of stenographic computer-assisted systems with hearing-impaired adults as a simulation of ASR, I have listed below some long-term objectives for an advanced general purpose ASR system, for use by hearing-impaired people.

1. Process continuous speech at rates up to 200 words per minute
2. Possess a base vocabulary of 50,000 words, plus space for the addition of 10,000 words
3. Conform to a speaker-independent configuration
4. Yield better than 95% verbatim accuracy
5. Produce text display within two seconds of spoken utterance
6. Contain a legible and convenient text display
8. Offer easy portability

In 1979, Houde, an expert in the area of ASR then and now, said:

For at least 40 years, the same answer has been given to the question: "When will automatic speech recognition be achieved?" The answer has always been "in ten years." At this time, there does not appear to be good reason to shorten this traditional estimate. (p. 572)

When I asked him to update his statement in 1993, he replied, "Let's make it five."

It is impossible to predict when all these pieces are likely to come together. Certainly, some will be ready before others, and these should be examined carefully with a view to targeted applications. Additional pieces, but not all, will probably come together and be in place within five years, ready to be used by people with hearing impairments in a number of applications.

CONCLUSION

In this chapter, we have discussed several approaches to the use of real-time speech-to-text communication by people with hearing impairments. However, of all these approaches, only handwriting and typing are interactive in nature (i.e., providing for both receptive and expressive communication on the part of hearing-impaired people). From the perspective of the hearing-impaired person, most systems provide only for reception.

As "hearing impaired" is generally defined, it can be said that most hearing-impaired people have intelligible speaking skills; we can assume that these people can literally "speak for themselves." For the majority, this is true, but our definition of hearing impaired also includes people who are deaf, many of whom are unable to speak intelligibly, and others who, for various reasons, are reluctant to do so. For these people, most systems we have described constitute only partial systems. For this reason, we must continue to give attention also to ways in which deaf people can become more active and "expressive" partners in communication with people who hear, to enable them to express themselves as full participants. Until this problem is solved, many deaf people will continue to rely on interpreters much more than on technology for interactive communication.

ACKNOWLEDGMENTS

I acknowledge friends and colleagues for their assistance in reviewing and suggesting changes and additions to this chapter. My thanks to Judy Brentano, Peter Jepsen, Joe Karlovits, and Linda Miller, all at the forefront of real-time captioning. I appreciate Bob Houde's sharing his expert knowledge and opinion about automatic speech recognition with me. For helping me seem to be more computer-literate than I am, I thank Don Beil and my son, Randy Stuckless.

REFERENCES

Abdelhamied, K., Waldron, M., and Fox, R. A. 1992. A two-step segmentation method for automatic recognition of speech of persons who are deaf. *Journal of Rehabilitation Research and Development* 29:45–56.

Blake, C. J. 1882. The glossograph: Automatic stenographic machine for the mechanical transcription of human speech. *American Journal of Otology* 4:190–93.

Conrad, R. 1979. *The Deaf School Child*. London: Harper & Row.

Fay, E. A. 1883. The glossograph. *American Annals of the Deaf* 28:67–69.

Feder, B. 1992. Different strokes for computing. *The New York Times* August 9, pp. F8–9.

Foster, S. 1992. *Working with Deaf People: Accessibility and Accommodation in the Workplace*. Springfield, IL: Charles C Thomas.

Gopher, D., and Raij, D. 1988. Typing with a two-hand chord keyboard: Will the QWERTY become obsolete? *IEEE Transactions on Systems, Man, & Cybernetics* 18:4:601–609.

Henderson, J., McKee, B., and Stinson, M. 1990. The NTID Computer-aided Speech to Print Transcription System (C-Print). Paper presented at the 17th International Congress on Education of the Deaf. Rochester, NY: National Technical Institute for the Deaf.

Hinton, D. 1992. *Examining Advanced Technologies for Benefits to Persons with Sensory Impairments*. Science Applications International Corporation report No. 92/1059 prepared for Office of Special Education Programs, OSERS, Dept. of Education, Washington, DC.

Houde, R. 1979. Prospects for automatic recognition of speech. *American Annals of the Deaf* 124:568-72.

Kanevsky, D., Nahamoo, D., Walls, T., and Levitt, H. 1992. Prospects for stenographic and semi-automatic relay service. Paper presented at meeting of International Federation of the Hard of Hearing, Tel Aviv, Israel.

Kurzweil, R. 1990. *The Age of Intelligent Machines*. Cambridge, MA: Massachusetts Institute of Technology.

Kyle, J. 1990. The HI-LINC project. In *Abstracts of presentations: Proceedings of the 17th International Congress on Education of the Deaf*, eds. R. Stuckless, D. Hicks, and G. Kingsford. Rochester, NY: National Technical Institute for the Deaf.

Lea, W. A. 1980. The value of speech recognition systems. In *Trends in Speech Recognition*, ed. W. A. Lea. Englewood Cliffs, NJ: Prentice-Hall.

Lee, K.-F. 1990. Context-dependent phonetic hidden Markov models for speaker-independent continuous speech recognition. *IEEE Transactions on Acoustics, Speech, and Signal Processing* 38:599–609.

Makhoul, J., Jelinek, F., Rabiner, L., Weinstein, C., and Zue, V. 1989. White paper on spoken language systems. In *Proceedings of the Speech and Natural Language Workshop*, DARPA. San Mateo, CA: Morgan Kaufmann Publishers.

McCrum, R., Cran, W., and MacNeil, R. 1986. *The Story of Language*. NY: Viking.

Miller, L. 1989. What is real-time captioning and how can I use it? *SHHH Journal* 10:7–9.

Roe, D. B., Mikkilineni, R. P., Prezas, D. P., and Wilpon, J. G. 1991. AT&T's speech recognition in the telephone network. *Speech Technology* 11:16–21.

Schein, J. D. 1968. *The Deaf Community*. Washington, DC: Gallaudet University Press.

Schein, J. D., and Delk, M. T. 1974. *The Deaf Population of the United States*. Silver Spring, MD: National Association of the Deaf.

Stinson, M., Stuckless, R., Henderson, J., and Miller, L. 1988. Perceptions of hearing-impaired college students toward real-time speech to print: RTGD and other educational support services. *Volta Review* 90:341–47.

Stuckless, R. 1978. Technology and the visual processing of verbal information by deaf people. *American Annals of the Deaf* 123:630–36.

Stuckless, R. 1982a. *Deaf and Hearing Students' Ability to Detect and Correct Word Errors in the Real-time Graphic Display of Spoken Lectures*. RTGD working paper No. 2. Rochester, NY: National Technical Institute for the Deaf.

Stuckless, R. 1982b. *Real Time and Delay in the Steno/Computer Transliteration of Live Spoken Lectures into Graphic Display*. RTGD working paper No. 3. Rochester, NY: National Technical Institute for the Deaf.

Stuckless, R. 1983. Real-time transliteration of speech into print for hearing-impaired students in regular classes. *American Annals of the Deaf* 128:619–24.

Stuckless, R., and Matter, J. 1982. *Accuracy in Transliteration of Spoken Lectures into Real-time Graphic Display for Deaf Students: First Year*. RTGD working paper No. 4. Rochester, NY: National Technical Institute for the Deaf.

Turn, R. 1974. *The Use of Speech for Man-machine Communication*. RAND Report-1386-ARPA. Santa Monica, CA: RAND Corp.

Virvan, B. 1991. You don't have to hate meetings—Try computer-assisted notetaking. *SHHH Journal* 12:25–28.

Waibel, A., and Lee, K. F. 1990. Why study speech recognition? In *Readings in Speech Recognition*, eds. A. Waibel and K. F. Lee. San Mateo, CA: Morgan Kaufmann Publishers.

Williams, J. 1987. Covert linguistic behavior during writing tasks: Psycho-physiological differences between above-average and below-average writers. *Written Communication* 4:310–28.

Wilson, D. L. 1992. Dramatic breakthroughs for deaf students. *The Chronicle of Higher Education* 15:A16–17.

Woodcock, K. 1992. Print interpreting for deafened adults. *1992 ALDA Reader*. Chicago: Association of Late Deafened Adults. (P.O. Box 641763, Chicago, IL 60664-1763).

Chapter • 12

VISUAL ALERTING AND SIGNALLING DEVICES

Carl J. Jensema and Debra L. Lennox

Sound plays an important function in alerting people who can hear: the footsteps of an approaching person, the ringing of a doorbell or telephone, and other unique, sound-producing events. If a person is hearing impaired, the sound must either be enhanced to a level that can be heard or conveyed to the person through some other sense. This chapter discusses how hearing-impaired people can be made aware of environmental sounds through the application of technology.

Few hearing people have ever stopped to think of the role environmental sounds play in their lives. Each of these many sounds provides a unique bit of information. This information is raw data that the human mind analyzes. The conclusions a person may draw concerning a particular sound or sequence of sounds may vary tremendously. For example, the sound of an opening door may signal the arrival of a friend, or it may indicate an intruder. The important point is that hearing people have this raw data to analyze and hearing-impaired people do not.

The challenge is to use technology to help hearing-impaired people obtain information that hearing people obtain through environmental sound. This is not a trivial task. Technology can provide assistance, but it cannot completely bridge the gap. The degree of assistance that technology can provide depends on a number of factors; the three major ones are the availability of appropriate technology, creativity in application, and availability of funding.

The driving force behind all of this is need. Certainly, hearing-impaired people do not need or want to be aware of every possible environmental sound. For every hearing-impaired individual, the need to know about a specific environmental sound has a particular value. A person may place no value on the need to know if his or her spouse snores at night. That same person may place great value on knowing if a smoke detector has been activated. The values vary with the person and the circumstances, but there are certain sounds, or categories of sounds, that hearing-impaired people, in general, want to know about. These include telephones, doors, babies, alarm clocks, pagers, fire alarms, and security alarms.

After it is decided *what* a person needs to be aware of, we must decide *how* to make the person aware. Hearing-impaired people can be alerted to environmental sounds generally in three ways: (1) enhancement of the sound to a level they can hear; (2) converting the sound to a visual signal; and (3) converting the sound to a tactile signal.

The first method, enhancing the sound, is often the most economical approach, provided the hearing-impaired person has some usable hearing. For example, it is often a simple matter to equip a doorbell with a bell that makes a louder noise. The main drawback is that enhancing the level of loudness to a degree where a hearing-impaired person can hear the sound may make it annoying to other people without a hearing loss.

The second method, converting the sound to a visual signal, is the most common approach. Light signals are silent, can alert people in a reasonably large area, and are usually relatively inexpensive. In this chapter, we emphasize devices that use light as a signal.

The third method, vibration, has some serious drawbacks that limit its applicability. In contrast to sound and light, vibration requires direct contact with the signal source. Many people dislike having a vibrating device attached to their bodies. The alternative, making a floor, chair, or some other appliance vibrate, can be both expensive and impractical. To be portable, body vibrators require battery power, and batteries rapidly weaken because of the power drain needed to vibrate a mechanical object. As a result, the use of vibrators has been limited mostly to bed vibrators that use 110-volt house current and to pagers that are required to vibrate only occasionally. Moreover, human skin loses its sensitivity to pressure when the pressure is applied repeatedly. For example, most people do not really "feel" the wrist watch they wear. A vibrating wrist band also tends to be ignored if it vibrates almost constantly over a period of time.

Finally, no introduction to alerting systems would be complete without stating the cardinal rule: "Keep it simple, reliable, and economical." For example, if you need to know if someone enters the door of your office, consider turning your desk to face the door, or investigate a strategically placed mirror costing $1.98 before you spend $300 on a customized electronic motion detector. There are many simple, low-tech solutions that can fit specific situations.

Some common categories of environmental sounds that hearing people need to be aware of and some solutions that may be applied need to be discussed. Many solutions depend on a relay switch. A relay switch uses electrical power in one circuit to turn on the power in another circuit. For example, one circuit may have 12 volts and another may have 115 volts. When the 12 volt circuit is closed, it creates an electromagnetic field in the relay that causes the switch of the 115-volt circuit to close. This may turn on a 115-volt lamp or start almost any 115-volt electrical device. The main advantage of using a relay switch is that it protects users from high voltage shock.

TELEPHONES

Thanks to a variety of new services, telephones now play a much larger role in the lives of hearing-impaired people than they did only a few years ago. Before these services were available, deaf people were limited mostly to using a TDD (telecommunications devices for deaf people) to call other people with TDDs. Or, he or she could call by voice to those people whom they could understand over the telephone (see Chapter 8, this volume). New services have greatly broadened the capability of all hearing-impaired people to communicate by telephone.

Unfortunately, these services do nothing to make hearing-impaired people aware of a ringing telephone. Devices are still needed to alert a hearing-impaired person to incoming calls. These devices can be divided into five groups: (1) sound enhancers; (2) line-powered strobes; (3) wired strobes; (4) wired incandescent devices; and (5) wireless transmitters.

Sound Enhancers

Several plug-in devices provide a loud telephone ring in such noisy environments as factories. These can also be used by hard of hearing people. The largest of these devices can emit a horn or warble sound of up to 105 dB. Fortunately, most of these devices have a volume control. They may be useful in some circumstances where the sound will not annoy hearing people. Setup is easy: simply plug the device into a telephone line. Most use the 97-volt ring surge that goes through a telephone line to signal an incoming call. They range from $20 for the small units, to $130 for the industrial ear-busters.

Line-powered Strobes

Line-powered strobes are among the easiest telephone alerting devices to hook up. These are small strobe lights that, as the sound enhancers mentioned above, use the 97-volt ring signal as a power source. Their usefulness is somewhat limited by their relatively low light level and the need to place them near the telephone because of limited cord length. They can be difficult to see in a large, well-lit room unless a person happens to be looking directly at them. On the other hand, they are compact and easy to use, making them ideal for travelers. They vary from $15 to $60, and they are available at many specialty stores.

Wired Strobes

Wired strobes are much like line-powered strobes, although they depend on 115-volt house power, rather than ring signal power. The ring signal is used to activate a relay switch within the device, which turns on a strobe light powered by house current. These units are usually brighter than line-powered strobes. Their greater complexity usually means they cost more. They range from $30 to $225. Some hardware stores, telephone specialty stores, and electronic specialty stores (e.g., Radio Shack) carry these telephone strobe lights.

Wired Incandescent Devices

Wired incandescent devices are, by far, the most popular visual method of alerting people to a telephone ring signal. These devices use the telephone ring signal to activate a relay switch that turns on an ordinary table lamp or some other 115-volt signaller. The device is usually a small box that plugs into a wall socket; the telephone line and a lamp are plugged into this box. The Fone Flasher (see figure 1) is an example of this device. These are probably the most popular telephone ring signallers on the market, and many thousands have been sold through hardware stores, telephone specialty stores, and electronic stores, especially Radio Shack. They sell for about $20 or less. For a single

Figure 1.   Fone Flasher. (Used with permission from Radio Shack.)

room and a single telephone, a Fone Flasher unit is usually the most economic and practical way to alert a hearing-impaired person to a ringing telephone.

There are also more sophisticated (and expensive) versions of the basic unit described above. They work on the same principle as the Fone Flasher, but they may have sturdier construction and a distinctive flashing pattern. They range up to about $100 at electronic specialty stores.

Two different types of devices are available that perform the same function as the ring voltage devices. One is a small box placed next to a telephone that is activated by the electromagnetic field produced when an older, rotary dial telephone rings. When activated, the box turns on a lamp or vibrator plugged into it. The drawback is that this device works only with older telephones. Newer electronic telephones do not produce the electromagnetic field necessary to activate the device. Because these telephones are disappearing rapidly, this device probably will not be available much longer. The second device is a sound sensor placed next to a telephone, which is activated by the sound of the ring rather than the electromagnetic field. A small box placed next to the telephone detects the ring and turns on a light or vibrator. The problem is that many of these devices are susceptible to false alarms from ambient room noise.

Wireless Transmitters

Finally, there are some sophisticated units that are activated by the telephone ring voltage and transmit a wireless signal to a remote receiver that turns on a light or vibrator. As many receivers as needed usually can be added to the system. The TR55 made by Sonic Alert is probably the most popular of these units. A transmitter and receiver cost about $80, with additional receivers costing about $35 each.

A variation of this is a sensor that transmits a signal to a central unit, which relays its own signal to remote receivers. In this case, the telephone sensor is only part of a broader master system of multiple sensors.

DOORS

There are four basic ways that a hearing-impaired person can be made aware of someone at the door. There can be sensors that detect the pushed doorbell button, the sound of a knock on the door, the opened door, or any movement near the door. The chosen method depends on circumstances and personal preference. For example, for a home situation, most people would probably prefer a flashing light when the doorbell rings. On the other hand, some apartments may not have doorbells and the dweller may prefer a sensor that detects a knock on the door. Alternately, there may be a situation wherein a hearing-impaired person works in a store or office where people enter without knocking, and it may be appropriate to have a light that flashes whenever a door is opened, or when someone walks through the doorway.

Doorbells

Doorbells are relatively simple devices that have three components connected by a small wire: a transformer that converts house current to low (8 to 16 volt) AC current, a button switch usually mounted near a door, and a bell. For a hard of hearing person, it may be enough to just add an extra bell or to replace an existing bell with a louder one. Many hardware stores carry bells, buzzers, or other items that may be suitable for the task. A simple bell that can be added to the existing doorbell may cost as little as $5.

For a deaf person, a louder bell is not enough. The solution is to add a device that operates in a manner similar to the telephone Fone Flasher. A small box wired to the doorbell system contains a relay switch that turns on a 115-volt lamp or vibrator. When the doorbell is pressed, the same current that activates the bell will close a relay switch and turn on a lamp. There are variations of this device, such as the one that contains a transmitter that activates a remote receiver, but the basic idea of having the doorbell power control a relay switch remains the same.

One variation deserves special mention. There are doorbell signalling devices that are intended to work in places where there is no conventional doorbell. For example, a deaf person living in an apartment without a doorbell may want to add a button at the door that flashes a light inside. The device contains all the components of a regular doorbell, except for the bell itself. It plugs into a wall outlet, a small wire is run outside the door, and a doorbell button is attached. When the button is pushed, a relay in the device closes and turns on a lamp or vibrator.

Doorbell devices for deaf people are less commonly available than telephone signal devices. They can be obtained from local dealers who specialize in devices for hearing-impaired people, or from a manufacturer of such devices. Sonic Alert and Phone TTY (see figure 2) are the best known. Prices for a doorbell system vary from about $75 to $125 for a single location. The cost will be higher if an alerting signal is needed at multiple locations.

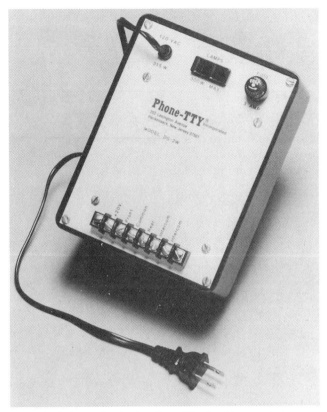

Figure 2.   Doorbell device by Sonic Alert. (Used with permission from Sonic Alert.)

Door Knockers

There are many situations where there is no doorbell and people customarily knock to get admitted. Over the years, various devices have been designed to alert a deaf person to a door knock. Most of these have had only limited success. Some are simply general purpose sound detectors hung on a doorknob. Others are more complex devices designed to detect only the specific sound of a knock on a door. Both have problems.

General purpose sound detectors are often activated by random room noise. Devices designed specifically to detect door knocks are an improvement, but not all doors are made of the same material and knocks on different doors produce different sounds. Both door knock alerting systems have reliability problems. It is recommended that a doorbell system be used instead of a door knock system whenever possible.

Of course, there are some situations where doorbells are not practical. For example, a traveler may want a portable door knock device for use in hotel rooms. Sonic Alert, Phone TTY, Ultratec, and others manufacture sound detection devices that can be used to detect door knocks.

Door-opening Switches

Most hearing-impaired people cannot hear a door being opened. There are many situations where this can be a problem, either for security or simply for peace of mind. For example, a family with young children may want a door-opening alert device simply to help keep track of the children's comings and goings. Or a hearing-impaired person working in a store or shop may want to be alerted of the entrance of a customer.

A number of inexpensive home security devices on the market can be adapted for such situations. These devices are usually designed to activate when a connection is broken. Half of an electromagnetic switch is mounted on the door, the other half is mounted on the door jamb. When the door is opened a connection is broken and the device is activated. Most such security devices have a 12-volt output that can be used to activate a buzzer, horn, siren, or a small 12-volt light. Should a person choose this device, he or she should be sure the device activation time is adjustable. If the device does not reset itself within a few seconds after the door is closed, it may not be suitable.

If a 115-volt lamp is to be used for the signal, a better approach is to go to a deaf-specialty dealer and purchase a doorbell system designed for use where there is no existing doorbell. Replace the doorbell button on this system with a magnetic switch on the door.

Motion Detectors

A few years ago motion detectors were expensive devices, but their cost has dropped dramatically. Hardware stores now commonly carry motion detectors costing as little as $15. There are two basic kinds of motion detectors. One sends out a beam that bounces back to the device via a reflector. When the beam is broken, the device is activated. These are good for situations in which a person wants to be alerted to movement at a specific point only. They tend to be more portable, with several free-standing units on the market that can be plugged into a wall outlet.

The second kind of motion detector emits a cone-shaped beam, and is activated whenever there is movement within that cone. These are often less expensive, but they are usually intended to be wired permanently into a building's electric system.

The use of motion detectors is limited only by creativity. For almost any circumstance that a hearing-impaired person should be alerted to motion, there is a way to apply a motion detector. For example, one deaf person had an L-shaped office with a desk out of sight of the doorway. For security reasons, he needed to know when people entered the office. A beam-type motion detector aimed across the doorway and wired to a 12-volt strobe on his desk solved the problem. The whole thing cost less than $100.

BABIES

Babies cry to communicate and there are many situations in which a hearing-impaired person cannot hear a baby's cry. Years ago, before electronic alerting devices were available, many deaf parents slept with their baby between them so the baby's movement during the night (rather than its crying) would awaken them. The risk of a parent rolling over and accidentally smothering the baby made this practice hazardous, but was often the only available solution.

Today the situation is quite different. A number of electronic gadgets are available to help hearing-impaired people become aware of situations in which a baby needs attention. The three main devices are sound detectors, motion detectors, and video cameras.

Sound Detectors

Many sound detectors are available for use as security devices or simply as conveniences. They generally turn on a light or whatever is plugged into them when a sound is detected. The sensitivity is usually adjustable. These may possibly be useful for monitoring a baby, but the devices turn on when *any* sound is detected. A perfectly happy baby may turn on the alerting system by shaking a rattle.

Special baby-cry sensors available from Sonic Alert, Phone TTY (see figure 3), and other manufacturers and specialty dealers are better. These sound detectors are designed specifically to be activated only by the sound frequencies of a crying baby. They will not be activated by adults talking, dropping toys, and similar noises. Sensitivity is usually adjustable to fit an individual baby. Most of these devices are wireless and consist of two boxes. One is a sound detector/transmitter plugged into the wall socket in a baby's room. The other is a receiver that is plugged into a wall socket in the parent's bedroom or whatever room in the house in which the signal is most needed. A lamp or vibrator can be plugged into the receiver. Prices are likely to be around $75 or $100 for a transmitter-receiver set.

Figure 3.   Baby cry sensor by Sonic Alert. (Used with permission from Sonic Alert.)

Motion Detectors

Motion detectors can be used with babies in some circumstances, but they are far less applicable and less reliable for this task than sound detectors. For one thing, many young babies may not move enough to set off a motion detector, even when they are screaming at the top of their lungs. A motion detector is most likely to be useful for older babies. For example, it could be used to alert the parent when a baby or young child becomes too adventuresome and begins to explore an off-limits area.

No motion detectors are commercially available that are specifically for baby or child monitoring.

Video Cameras

In the last few years the price of video equipment has dropped to a more affordable level. For example, a major toy store chain carries a black-and-white video camera and television monitor that sells for slightly over $100 and is intended for use as a toy, but it can be used for monitoring a baby. A higher quality set for security purposes sells for $300.

This kind of equipment is especially useful for baby monitoring when used in conjunction with a baby-cry detector. A parent can have a baby-cry device set up to awaken him or her at night and have a closed-circuit television monitor set up so the baby can be observed without the parent getting out of bed.

ALARM CLOCKS

Hearing-impaired people are usually unable to use an alarm clock with an audio signal. There are a number of alarm clocks that have a switched plug for use with a lamp or vibrator. Many of these are ordinary alarm clocks with the plug wired in by a specialty dealer. A few are specifically manufactured with the switched plug. The user simply plugs in a lamp or vibrator and sets the alarm. These devices often sell for $50 to $75, with the relatively high price reflecting the labor cost in adding the switched plug. The market is limited, and adding the plug effectively doubles the price of the clock.

There are several battery-operated travel alarm clocks available. These devices are roughly the size of a pack of cigarettes and contain a small digital alarm clock, a battery, and a vibrator. The user sets the alarm and puts the device in a pajama pocket or under their pillow before going to sleep. Prices are in the $30 to $50 range, and they are available from specialty dealers.

Many hearing-impaired people find it much cheaper and easier simply to go to a store that carries clocks and select a timer that is intended to turn lights on and off at preset times. These timers cost as little as $4 and may fill the needs of someone who must wake up at the same time each day.

PAGERS

A simple shout is often enough to get the attention of a hearing person, but paging a hearing-impaired person is considerably more difficult. Fortunately, a number of devices are on the market that use light or vibration and are adaptable to the needs of hearing-impaired people. A few are designed specifically for use by hearing-impaired people, but most are general purpose remote switches that may or may not have been modified. Two common situations must be considered: (1) mobile sender and stationary receiver; and (2) stationary sender and mobile receiver. Each requires different equipment.

Mobile Sender, Stationary Receiver

Sometimes a person needs a battery powered transmitter that can be used to activate a stationary receiver. For example, a person working at different sites around a warehouse may need to alert a central office that he or she needs assistance. Silent Call and Sonic Alert, among others, have pagers specifically designed for use in alerting hearing-impaired people. In addition, most stores that carry security devices have wireless, battery-operated transmitters that transmit a signal, which makes a remote receiver turn on a 115-volt lamp or other device. These often can be used as pagers for hearing-impaired people. Some specialty dealers carry versions of these devices that have been modified to make the receiving unit flash instead of merely turn on. Of course, such modifications add to the cost of the unit.

The important thing is to make sure the unit will work in the circumstances under which it will be used. Transmission distance varies, as does the ability to transmit through walls. Some units are limited to line-of-sight operation in a single room, whereas others may effectively transmit up to 100 yards and may transmit through walls. The cost of this equipment varies from $20 to $100, depending on power, quality, and other factors.

Stationary Sender, Mobile Receiver

In this case, a stationary sender is paging a hearing-impaired person who is moving around. In a reverse of the previous example, a central office might want to page a hearing-impaired person who is moving around a warehouse. There are two possible ways the hearing-impaired person could be alerted. Receivers with lights can be installed wherever the hearing-impaired person is likely to go or the hearing-impaired person can carry a battery-operated receiver with a vibrator. The first option, lights wherever a hearing-impaired person might go, is usually too complicated and expensive if the person moves around much. The second option, having the person carry a battery-operated vibrator, is usually the most feasible. Silent Call and Quest Electronics both make special vibrating devices for this situation. There are also many vibrating telephone pagers on the market that can be used. The disadvantage of these pagers is that a telephone must be used to activate them, a process that can be much more complicated than simply pressing a single button.

The Silent Call and Quest units cost several hundred dollars and have a limited range of operation. The telephone pagers have effective ranges of many miles and cost $15 per month and up.

FIRE AND SMOKE ALARMS

Fire poses a special danger for hearing-impaired people because most fire and smoke alarms use audio warnings which they may not hear. Hard of hearing people should test their smoke detectors to determine if they can hear them clearly *without* a hearing aid. Too many forget that they take off their hearing aids at night and cannot be awakened by ordinary smoke detector alarms. Others fail to realize that during the day background noise can make it difficult for them to hear a smoke detector alarm. The safest course of action is to obtain a special smoke alarm that provides both audio and visual alerts.

In the past, smoke and fire alarms have used audio warnings only. In recent years, there has been an effort to provide visual warnings. The Americans with Disabilities Act has provided a strong impetus for this. Regulations proposed in the Federal Register on September 6, 1991, page 45695, state that:

> Alarm signal appliances shall be integrated into the building or facility alarm system. If single station audible alarms are provided, then single station visual alarms shall be provided. Visual alarm signals shall have the following minimum photometric and location features:
>
> 1.   The lamp shall be xenon strobe type or equivalent.
>
> 2.   The color shall be clear or nominal white (i.e., unfiltered or clear filtered white light).
>
> 3.   The maximum pulse duration shall be two-tenths of one second (0.2 sec) with a maximum duty cycle of 40 percent. The pulse duration is defined as the time interval between initial and final points of 10 percent of maximum signal.
>
> 4.   The intensity shall be a minimum of 75 candela.
>
> 5.   The flash rate shall be a minimum of 1 Hz and a maximum of 3 Hz.
>
> 6.   The appliance shall be placed 80 inches (2030 mm) above the highest floor level within the space or 6 inches (152 mm) below the ceiling, whichever is lower.
>
> 7.   In general, no place in any room or space required to have a visual signal appliance shall be more than 50 feet (15 m) from the signal (in the horizontal plane). In large rooms and spaces exceeding 100 feet (30 m) across, without obstructions 6 feet (2 m) above the finish floor, such as auditoriums, devices may be placed around the perimeter, spaced a maximum 100 feet (30 m) apart, in lieu of suspending appliances from the ceiling.
>
> 8.   No place in common corridors or hallways in which visual signaling appliances are required shall be more than 50 feet (15 m) from the signal.

In simple English, this means that buildings falling under these federal regulations should have a fire alerting system with a very bright strobe light visible in every room and hallway. Privately owned homes are not subject to these regulations, but the general principal should still be followed and every home with a deaf or severely hard of hearing resident should have a visual smoke alarm.

Deaths from home fires are most likely to occur at night when people are sleeping. For this reason, sleeping areas should be given the highest priority in planning a fire warning system. The visual light system must be bright enough to awaken a sleeping deaf person; powerful white flashing strobe lights have been found to be the most effective for this. There are three particularly good portable smoke detectors on the market that have strobe lights rated at 100 candela or more and audio warnings rated at 85 dB or more. These are the Whelen (see figure 4), VenTek, and Gentex (see figure 5) strobe alarms. All operate on 115-volt house current and provide an excellent warning under most circumstances. A user can plug these units into a wall socket and hang them high on a wall near the ceiling. The Gentex smoke detector is also available as a permanent-mount model for wiring directly to a wall outlet box. Prices vary from $100 to $270 for this equipment.

Portable visual/audio smoke detectors may be adequate for a small home or apartment, but for large houses, a system is needed that will alert a person of a fire at a remote location. There are several options for this situation. First, Silent Call manufactures a special smoke detector that sends a wireless signal to a receiver. The receiver activates the Silent Call on-body vibrator, bed vibrator, or strobe attachment.

Second, some people attach a sound detector device to a regular smoke detector and have the sound detector turn on a light or vibrator. This is not recommended because general-purpose sound detectors are not intended for this purpose and may generate false alarms or may malfunction. A few manufacturers of specialty equipment recommend use of their sound detectors with regular smoke detectors. This may be acceptable, but the customer should be certain that they

Figure 4.   Strobe fire alarm by Whelen. (Used with permission from Whelen Engineering Company.)

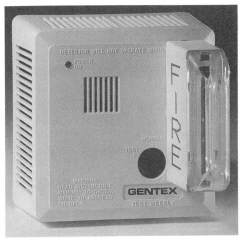

Figure 5.   Strobe fire alarm by Gentex. (Used with permission from Gentex Corp.)

will perform correctly. A potentially deadly fire situation is not the place to test a "Rube Goldberg" contraption. Stay with proved and approved devices for smoke detection and fire-warning systems.

Finally, the best fire protection system for a large house is a professionally installed whole-house custom-designed smoke detection system with provisions for visual alerting. A person who elects this system should look in the telephone yellow pages for security system installers and find one who will work with a homeowner to develop a custom system with appropriate signal lights and enhanced sound.

SECURITY ALARMS

The need for hearing-impaired people to be alerted to burglar alarms through the use of flashing lights has been somewhat neglected by the home security industry. This is a foolish business practice because flashing lights, in addition to an audio alarm, are likely to scare off an intruder as well as to alert a hearing-impaired resident.

On the other hand, most burglar alarm devices are designed to activate a 12-volt siren and the same 12-volt circuit can also be used with relay switches to turn on 115-volt lights or vibrators. Most burglar alarms can be adapted in this manner, but such an adaption usually requires knowledge of electric circuitry and should be done only by a qualified electrician. This is especially true with the newer electronic burglar alarms. A hearing-impaired person with all the major doors and windows in the home wired with a central burglar alarm, and who knows little about electronics should consult a specialist about adding alerting lights, rather than trying to be a do-it-yourselfer.

There are exceptions to the don't-touch rule. For example, such specialty manufacturers as Sonic Alert have devices that can be connected and used for home security purposes by hearing-impaired people with little technical knowledge. Another is the application of general-purpose, portable home security devices sold by many hardware and electronic stores. Radio Shack's portable beam-type motion detector has a 12-volt outlet that can be used with the Radio Shack 12-volt strobe light to warn of an intruder. This requires nothing more than a screwdriver and the ability to read simple directions.

CUSTOM-DESIGNED SYSTEMS

Most devices that have been discussed so far are more-or-less portable. When a person moves to another house or changes the design of an alerting system, it is a relatively simple matter to disconnect the devices and reinstall them in a new place or new configuration. A problem with portable systems that are physically wired is that they tend to have unsightly trailing wires. The drawback with portable systems that transmit through the air or through the electrical wiring of a house is that they are more prone to interference and false positive responses.

Custom-designed systems side-step these drawbacks. They are hardwired, but the wires are hidden from view. The whole system is designed to be unobtrusive. It's akin to comparing extension cords with regular house wiring. Both operate effectively, but one is temporary and economical, whereas the other is permanent and more costly. Actually, the difference in cost is not always great, if the life of the device is considered. Custom-designed systems are part of the house and, as the rest of the house circuitry, the system should last for many years. Custom-designed systems cost more because of the installation labor required, but over a system's 20- or 30-year life span, this higher cost may not be much.

The choice between a portable system and a permanent system often depends on how long the system is to be used. If a person rents an apartment for a year and plans to move at the end of that time, a permanently installed, custom-designed alerting system that cannot be easily removed makes little sense. However, if a person buys a house and plans to live in it for several years, a custom-designed system is the best choice.

A key part of getting a good custom-designed alerting system installed is finding the right installer. Many cities have electricians who specialize in alerting systems for hearing-impaired people. Anyone considering a custom-designed alerting system should ask in the local hearing-impaired community for an electrician who does this work. This is also a good way to check the reputation of an electrician you are considering. If an electrician with experience in custom-designed alerting systems for hearing-impaired people cannot be found, an electrician specializing in home security systems is the next best option.

Once an electrician has been selected, there will be a preliminary meeting to decide exactly what characteristics the alerting system will have. What sounds need to be detected? Where should signal lights be placed? Where and how will the wiring be installed?

This last question needs expansion. Many custom-designed systems cause ordinary room lights to flash as alerting signals. This requires the use of a low-voltage relay to turn room lights on or off. The relay can be located at a central control panel or in the room lightswitch. If the relay is at a central control panel, the wiring to the room switch must carry 110 volts. If the relay is located in the room lightswitch, the wire to the switch need only carry 12 volts (or whatever low voltage is being used).

Unless 110-volt lines can be conveniently routed through an attic or basement, it is often difficult to hide them. On the other hand, a 12-volt line needs only a small-gauge wire and is much easier to hide. A good electrician will be able to hide it behind baseboards, under rugs, in walls, through heat return ducts, and in other places.

Figure 6 is a basic diagram of how a standard house circuit can

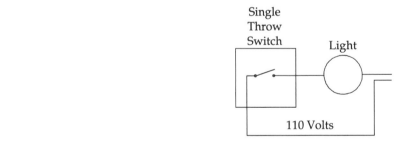

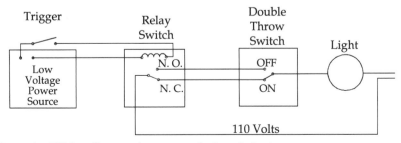

Figure 6. Wiring diagram for custom-designed alerting system.

be changed to accommodate a custom-designed alerting system. The top illustration shows a light turned on by a simple wall switch. The bottom illustration shows the addition of an alerting device. The two-way switch has been replaced by a three-way switch and one line to it is controlled by a relay. The relay is activated by a trigger, such as a doorbell button or some other switch. Relays come in different sizes and it is possible to obtain one small enough to fit in the wall switch box. In the system shown in figure 6, when the trigger is activated, the light goes off if it was on; and it goes on if it was off. It remains in this state until the trigger is deactivated.

Figure 6 is probably the simplest example of a custom-designed alerting system. With multiple triggers and many relay switches, it can become enormously complicated. For this reason, it is important to find an electrician who understands what needs to be done and has the technical skills to develop the system.

SUMMARY

This chapter has described some basic systems to provide awareness of the environmental cues used by hearing-impaired persons. Whether it be a simple, plug-in light that flashes when the phone rings or someone knocks on the door, or an elaborately wired signalling system that picks up all possible sounds, the assistance and independence that a device offers to a hearing-impaired person is invaluable. The products described are applicable both at home and at work.

With the passage of the Americans with Disabilities Act in January 1992, increased attention has been focused on the needs of hearing-impaired persons at home, at the office, and in public areas. It is important to remember that, although a hearing person may not want to be aware of every environmental sound at a particular time, for the hearing-impaired person, not having these noise cues could have untoward consequences. It is a simple matter to provide awareness for important sounds and the results can be of immeasurable benefit.

For more information on the products mentioned in this chapter, contact the manufacturers listed in the appendix, or consult your local electronics specialty store.

APPENDIX: RESOURCES

Gentex Corporation
10985 Chicago Drive
Zeeland, MI 49464
(616) 772-1800

Phone-TTY, Inc.
202 Lexington Ave.
Hackensack, NJ 07601
(201) 489-7889

Quest Electronics
510 South Worthington Street
Oconomowoc, WI 53066

Radio Shack Division of Tandy
Corporation
300 One Tandy Center
Fort Worth, TX 76102
(817) 390-3400

Silent Call Corporation
PO Box 16348
Clarkston, MI 48016
(313) 673-6069 TDD

Sonic Alert, Inc.
1750 W. Hamlin Road
Rochester Hills, MI 48309
(313) 656-3110

Ultratec
6442 Normandy Lane
Madison, WI 53719
(608) 273-0707

VenTek, Inc.
PO Box 130608
Birmingham, AL 35213
(205) 967-9127

Whelen Engineering Co.
Rt. 145 Winthrop Rd.
Chester, CT 06412-0684
(203) 526-9504

Special Populations

Chapter • 13

Combined Hearing and Vision Impairments

Mitchel B. Turbin

The effects of combined hearing and vision loss can range from modest inconveniences to near catastrophic barriers that can result in mind-numbing frustration and isolation. Similarly, the "reasonable" accomodations to ensure the optimal access guaranteed by the Americans with Disabilities Act can range from simplicity to complexity. The former set of conditions can be illustrated with the example of a hearing- and sight-impaired individual who uses an amplified telephone that also incorporates oversized numerals on the keypad and costs less than $100; this person may perform nearly all life functions independently with the assistance of some technologic and environmental accomodations that are neither exotic nor costly. The latter case, however, is typified by the totally deaf and blind person. Whereas such an individual can achieve a rich and full life, he or she may have to depend to a considerable degree on the intervention of interpreters and sighted guides, as well as on a sophisticated array of expensive mechanical and electronic aids. A basic, single-line residential Braille-text telephone system can easily cost as much as $5,000.

The limits caused by physical and sensory disabilities to certain aspects of human life are exacerbated for people with combined sensory loss because *vision* is the sense most used to compensate for hearing loss, and *hearing* serves the same function for most sight-impaired people. Impairment in both senses truly is the proverbial trap between a rock and a hard place. Thus, the rule of thumb commonly cited is probably an accurate portrayal of the proportions of the problem: combining hearing to vision impairments does not result in *adding* the handicapping effects of one to the other, but rather *multiplying* them.

Nevertheless, I stress again that with proper accomodations, hearing- and vision-impaired people, even totally deaf and blind people, can achieve remarkable accomplishments and make valuable contributions to society. Sight- and hearing-impaired Americans have made their marks in white and blue collar vocations, academe, the arts, community affairs, social services, and the sciences. Although this chapter cannot possibly address all the dimensions of providing access for this diverse group of people, it does supplement the other chapters in this volume, and it provides an introduction to the nuts and bolts of access for this special population. It should be noted that I deliberately do not use any single label—deaf-blind, blind-deaf, or dual sensory impaired. I feel it is imperative that we do not lose sight of the diversity—the individual differences—within this group of human beings. My shifting terminology may prove irksome for those readers accustomed to the tendency in the professional literature to demand consistency of terminology. However, the population addressed here is not monolithically consistent; I hope complex language will allow this human complexity to imprint itself upon the reader's mind.

POPULATION CHARACTERISTICS, ETIOLOGY, AND BASIC IMPLICATIONS

How many hearing- and sight-impaired people live now in the United States? A 1982 study funded by the United States Rehabilitation Services Administration estimated that the noninstitutionalized population of people with combined hearing and vision loss was over 734 thousand. Of these, about 42 thousand were considered "deaf-blind"—in other words, lacking *effective* use of both hearing and vision. Another 25 thousand were listed as "deaf and severely visually impaired." Thus, we have almost 70 thousand people who, as we discuss later, require some form of interpreter services for many aspects of communication access. However, most of the individuals assessed in this study were those who were severely hard of hearing: 358 thousand of whom were also legally blind, and 309 thousand of whom were considered severely visually impaired.

Before we can discuss actual methods of providing access for our target population, we must understand some of the pertinent features of visual impairment and of visual functioning. Certainly, as in hearing loss, we see here the full gamut of sensory degradation, from total inability to perceive any light, all the way to various degrees of noncorrectible near-sightedness. The two-part definition of legal blindness gives us a basic foundation for grasping those aspects of visual impairment most frequently of concern to people who also have hearing loss.

One definition of legal blindness is: best corrected *visual acuity* of less than 20/200. This means that a person can not discern details at a distance of 20 feet (even with properly prescribed corrective lenses) that a person with "average" vision would be able to discern from 200 feet away; the same ratio holds true if the distances are, for example, 2 feet and 20 feet, respectively. The second definition of legal blindness involves *visual field*: a "normal" visual field is about 160°, a visual field of 20° or less, commonly called "tunnel vision," denotes legal blindness.

Although either or both of these conditions may be present, an individual may have normal functioning in some circumstances, but have serious constraints in others. Thus, a person with an impairment of acuity may be unable to read normal-sized type, but can walk about freely, even in the dark. There are, for example, people with macular degeneration, a retinal condition in which central vision is severely affected, whose peripheral vision, which is also a primary feature of night perception, is unaffected. Another person with "tunnel vision" may have no difficulty doing fine, detailed work, but may have to use a guide dog for traveling. Various environmental conditions, including lighting, color contrast, and fatigue, can also affect most people's visual capability, and there are many individuals whose vision is compromised in a variety of ways simultaneously.

Finally, brief attention to the etiology of hearing and vision loss is useful. One frequently encountered condition within this population is Usher syndrome. The rehabilitation literature has noted repeatedly that as many as half the people *seeking rehabilitation services* for combined sensory loss have this syndrome, although it is clear and should be noted that they do not represent anything near that percentage of the total population cited in the REDEX study (thus, we can see that many who need and deserve services traditionally have *neither* sought nor received them). Usher syndrome is characterized by congenital hearing impairment, in which people are either profoundly deaf (Usher syndrome type 1) or hard of hearing (Usher syndrome type 2); and progressive blindness caused by retinitis pigmentosa (RP). RP generally is first noticed in late adolescence or early adulthood by the telltale signs of night blindness and tunnel vision. Night blindness usually becomes increasingly severe, and the visual field almost always continues to constrict into legal blindness, becoming severely restricted in most people by midlife. Total blindness does develop in some individuals, although others retain useful central vision until late in life. Cataracts—a condition also often found in the general population—are the growth of scar tissue that eventually occludes the eye's natural lens, and are also quite common among those with RP.

Finally, it is extremely important to realize that hearing and vision loss can occur at any time in life and from a great variety of causes, including maternal rubella, diabetes, meningitis, and trauma. People may have RP, macular degeneration, glaucoma or any other visual impairment, and develop a hearing loss for totally unrelated reasons. On the other hand, someone might have been born deaf or hard of hearing, and develop a medically unrelated vision impairment. The point in a person's life at which the combined sensory loss becomes significantly impairing, and the order—whether sight or hearing are impaired first, if the two do not become impaired simulaneously—are major considerations in determining the kind of accomodations that will be required for communication access.

Those who have been deaf from childhood tend to have attended special education programs, probably have learned sign language, and generally are socialized within the deaf subculture. Although they make up a minority of our population of concern here, this is the group—particularly those with Usher syndrome type 1—who form the majority of the people who join local, state, and national organizations, including the American Association of the Deaf-Blind. Accommodations for these people often center on American Sign Language, with modifications to compensate for their limited or missing vision. The most remarkable modification is the presentation of sign language, not to a person's eyes, but into their hands. I have repeatedly encountered adult sign language users (deaf culture members) who had a relatively easy time making the transition to a form of nonvisual, tactile signing when their vision declined. Clearly, their lifelong experience of signing is transferable to both expressive and receptive skills in tactile signing. However, many people prefer to utilize residual vision until little or none remains, and there are varying degrees of skill, both in using residual vision and in developing comprehension of tactile signing.

The so-called hearing culture subgroup tends to have different needs. Having had hearing in childhood and perhaps well into adulthood, they acquired relatively normal use of spoken and written English. Those deafened as adults, even with normal vision, infrequently learn American Sign Language (ASL), tending instead to use one or another English-based form of manual communication. Unfortunately, no true standard for this type of manual communication exists yet, even for sighted deaf persons. Some vision-impaired people, therefore, may rely on fingerspelling (made famous by Helen Keller—every letter is formed by the interpreter into the client's hand; effective, but both slow and fatiguing), whereas others incorporate varying amounts of ASL vocabulary and grammar. Here, also, it is important that interpreters individually tailor their communication styles to conform to those of their clients; we discuss this, and even greater alternative modalities below. It is interesting to note that even those among this group who are totally deaf usually continue to prefer to express themselves vocally.

None of these individuals, however, whether from hearing or deaf backgrounds, can keep up with the speed of ordinary spoken or signed communication for very long. Tactile signing, whether it approaches the grammar of ASL or of English, is necessarily much slower than spoken or visual signed language. It is more fatiguing for the disabled person to comprehend, and also more fatiguing for the interpreter to produce. The only solutions are to decelerate the communication, compress the information by transmitting main points only, and arrange for periodic rest breaks (important for both interpreter and client). Providing Brailled information in advance of what will later be signed often facilitates comprehension of signed communication.

On the other hand, many who have been deaf from childhood have poor literacy in English, even when sighted (the average reading level for an adult who grew up deaf in the United States is 3rd grade, although there are unquestionable exceptions to this). Therefore, they are unable to use even Brailled information well, and they are much more amenable to acquiring information when it is presented through the hands of an interpreter.

The question of when a vision impairment makes its impact felt is also important. For example, those who become significantly visually impaired early in life are likely to learn Braille, but this learning process becomes increasingly difficult as people age. Vision loss also takes its toll on people's independence by making travel increasingly difficult. Obviously, such symptoms as loss in central visual acuity, tunnel vision, and night blindness eventually make driving impossible and illegal. Moreover, walking and using public transportation also become major problems. One cannot use a bus if one cannot find the bus stop; identifying the correct bus route is another challenge when one can neither see the numbers, nor easily communicate with either drivers or fellow passengers.

It has been my experience that many factors can combine to determine whether a hearing- and sight-impaired person can get about on his or her own, or will become totally dependent upon others. I know totally deaf-blind people who travel extensively, and others, also bright, fairly young and healthy, perhaps with far fewer sight or hearing impairments, who have rarely left their homes at night for many years. People's temperaments, and whether they live or have lived in urban or suburban areas certainly are important factors, in addition to age at onset of visual loss, the rate of progression of the deteriorating vision, and their current age and state of health. It is also true that most guide dog schools do not train people who are impaired in both primary senses to work with canine assistance. We should understand the importance of using hearing as a travel aid for blind persons; hearing, for instance, can tell a blind person when a traffic light has changed, what direction traffic is coming from, how far away the traffic is, and (more or less), how fast traffic may be moving. And, if one can neither hear nor see, (and perhaps not even speak intelligibly, a common feature of childhood deafness) how does one ask for directions or even assistance to cross the street, or to find the bus stop, train station, or airport gate? There are ways, but not everyone has the energy or the chutzpah to use these strategies. The upshot of this is that providing access to certain events for certain people may also involve providing door to door transportation, and on-site sighted guides. Not for everyone, but for some.

COMPREHENSIVE ACCOMODATIONS: THE COMMUNITY MEETING

Because of the multifarious accessibility needs of the people we are concerned with in this chapter, I describe here the activities of an idealized community meeting. There are three precedents for this type of meeting. First, I have attended numerous national conventions of the American Association of the Deaf-Blind (AADB), which are held in different locations each year around the United States. These conventions are attended by 200 to 300 "delegates"—people with varying degrees of hearing and vision loss, or deaf-blindness, and an even greater number of volunteers who serve as interpreters, guides, and all-around assistants. The latter have come to be called support service providers (SSPs), and without them, AADB conventions could not take place. However, AADB conventions are organized by deaf-blind people for deaf-blind people—they are extraordinarily complex events, but they are also extraordinarily *empowering* events, and their successful annual occurrence over the past fifteen years is testament to the capabilities of the consumers, as well as the SSPs.

Second, I have participated in many meetings and social events in Seattle of the Washington State Deaf-Blind Citizens (WSDBC) organization. Occurring three or more times each year for over ten years, these are small AADBs. Most members of this educational-advocacy-social organization have Usher syndrome type 1, although there are also other participants (I have Usher syndrome type 2, for example).

Third, Metamorphosis is a support group for "English speaking, vision- and hearing-impaired people" that has met, more or less monthly, for seven years, also in Seattle. Most of the 8 to 15 members of this group have Usher syndrome type 2. Access needs for members of this group are typically quite different from those of WSDBC.

Seattle, with its large contingent of Usher syndrome and other similarly sight and hearing disabled consumers, has been the site of at least three large community meetings in recent years that were organized and funded by an ad hoc consortium of government groups concerned with service needs of this special population. Certain unique features of these meetings warrant mention. Finally, I have participated in various other meetings, cultural events, political actions, and outdoor and indoor recreational activities involving people with combined sensory losses.

Probably, the first important lesson we can derive from these events is that an individual or committee should be selected well in advance of the occasion in order to coordinate the various arrangements to ensure access. At least three areas must be addressed.

1. Transportation to and from the event, and in certain instances (a convention, or a tour to a scenic or recreational site, for example), people to orient partially sighted, and guide severely visually impaired and blind individuals
2. Communication access, i.e., interpreters, assistive listening systems, real-time captioning, and so forth
3. Provisions to make printed material accessible: regular print, large print, and Braille (both the basic level 1 Braille and the advanced level 2 Braille, if possible).

Obviously, the coordinator must have advance information regarding individual needs of participants. Thus, an application/information form becomes essential to provide assessment. A sample of such a form is found in table I, modified from a form used for the 1993 AADB Convention, and several other forms from different activities. A somewhat similar form can be developed to obtain information on the skills of people who will serve as interpreters, guides, and so forth. Support service providers (SSPs) are people who serve as both interpreters and sighted guides, either volunteer or professional, for clients impaired in both vision and hearing. In some areas and circumstances, these workers are less skilled than certified interpreters and should not, if adequate funding is available, replace them. A general description of sign language interpretation is found in Chapter 10, this volume.

Working our way through the sample application form, we can compile a comprehensive view of most strategies for obtaining access for our target population. Items 1 and 2 are self-explanatory; items 3 and 4 are significant in the three end-of-line options: (V)oice, TTY (text telephone, or telephone device for the deaf), and (TB) telebraille—

Table I.   Sample Application Form for Hearing and Vision-Impaired Access

1.   _____
     Name: Last, First, Middle Initial

2.   _____
     Address: Street   City   State   Zip

3.   Day Telephone: Area Code + Number_____V__TTY__TB__

4.   Evening Telehone Number_____V__TTY__TB__

5.   Date of Birth_____   6.   Sex: Male__Female__

7.   My vision is best described as: Sighted ____ Blind ____
     Tunnel Vision ____Visually Impaired ____

8.   Media: Braille 1 _____   Regular Print _____   Audio Cassette_____

     Braille 2 _____   Large Print _____

9.   Mobility:

     I need a guide at all times _____

     I need a guide at night _____

     I need a guide on outside tours _____

     I don't need a guide _____

10. I plan to arrive (when? bus/train/plane/private car?):

     _____

11. I need help with transportation to the event: Yes _____   No _____

     What help? _____

12. I will bring a dog.   Yes _____   No _____

13. I will bring a wheelchair.   Yes _____   No _____

14. My hearing is best described as: Hearing _____   Deaf _____

                                             Hard of Hearing _____

15. If Hard of Hearing, I use:

     Hearing aid with T-switch _____

     Hearing aid without T-switch _____

     FM System _____

Audio Loop System _____

Voice shadow _____

Real-time captioning _____

Other: _____

16.   If use interpreting, I prefer:

Platform interpeter _____

Close visual interpreter (small group) _____

Close visual interpreter (one-to-one) _____

Tactile interpreter _____Receive lefthand _____ righthand _____

Tracking _____

Tellatouch (or other typing) _____

Other (please describe)_____

17.   Receptive language:

ASL _____PSE _____SEE _____Fingerspelling (American)_____

Fingerspelling (British) _____Tadoma _____ Oral Interpreter _____

Other _____

18.   I have an additional disability you should know about _____

19.   I prefer this SSP/Interpreter: _____

Name and telephone number (if known)

We will arrive together.   Yes _____   No_____

20.   Any special dietary needs? (vegetarian, diabetic, kosher, etc.)_____

which are Braille output phones for deaf-blind people. Note that individuals may check more than one of these items, reflecting the "borderline" use of residual hearing or vision, as well as the presence of family members or roommates.

The two basic questions, items 5 and 6, may be important because although most people are flexible in regard to working with SSPs of either gender and any age, others may be better matched with same sex and age workers. For example, if the person requires a guide to find his or her way to and within the bathroom, an SSP of the same sex obviously is preferred.

Items 7 and 8 relate specifically to vision loss. Categorizing people according to their particular type of loss may facilitate organizing services. Most people are generally familiar with Braille, the tactile system in which the alphabet, numbers, and punctuation marks are denoted by raised dots on paper, or mechanical, or electronic media. Braille 1 is a fairly straightforward translation of all material into the raised-dot Braille "cells"; because this results in quite bulky material, and the reading of tactile information is necessarily somewhat cumbersome, Braille 2 was developed, utilizing numerous word contractions to speed up reading and reduce the bulk of printed material. It cannot, however, automatically be assumed that all users are skilled in this more advanced system; if an either/or situation exists, Braille 1 is the safer choice. Most cities have Brailling services, either at the local Lighthouse for the Blind, Library for the Blind, or other volunteer or public agency. The state agency for vocational rehabilitation services for blind people can refer people to a Brailling service. As noted earlier, attention may also need to be paid to the literacy level of the consumers being targeted: accessibility may mean being certain that the language used is exceptionally clear and easy to understand, as well as available to whatever sensory capacity exists.

In this day of the ubiquitous word processor and desktop publishing software, rendering material into large print is relatively easy. Some people with vision impairment can do quite well with material printed in 12 point type size; 14 and especially 18 point are likely to be even more useful. Examples of these are included in table II. Using the boldface feature of typical word-processing software will also significantly increase readibility. Obviously, deaf and profoundly hearing-impaired people cannot use information presented on audio tapes. However, with good playback equipment and proper earphones (including hearing aids with direct audio input, or induction silhou-

Table II. Examples of Different Size Type Fonts

This is an example of 10 point type.

This is an example of 12 point type.

This is an example of 14 point type.

This is an example of 18 point type.

ettes, obtainable from qualified audiologists and hearing aid dealers) others with moderate and even severe losses may be able to receive information effectively in this manner.

BASICS OF MOBILITY AND SIGHTED GUIDE TECHNIQUE

Items 9 through 11 refer to mobility needs of those with impaired sight. Certainly there are many instances in which organizers of various events are unable to provide sighted guides, particularly to and from the event, and the consumer will have to rely upon his or her own resourcefulness and independence. Public mass transit is required to be accessible to all individuals with disabilities under the ADA; however, the strategies by which this requirement is met lie outside the scope of this volume. It should be clear, I hope, that even the most resourceful and independent consumers may face formidable challenges in independent travel, even where transit companies have made accomodation arrangements: where car pool or chartered vans or buses are a viable option, such a provision of services will greatly increase the likelihood of participation by consumers. Moreover, a basic understanding of sighted guide technique is an essential part of the repertoire of skills of an SSP, and many interpreters and others working with sight- and hearing-impaired clients should also be comfortable in this activity.

Figure 1 provides a simple demonstration of proper form (obviously, this is a skill best practiced with real people, not simply learned

Figure 1.   Using tactile sign language to provide environmental description to a deaf-blind man. (Photograph by Jerry Weissman.)

from a book). Note that the guide should place him or herself slightly ahead of the client. The client generally will find it optimal to grasp the guide's elbow lightly, and the guide will find that by moving his or her elbow up or down, or slightly behind his or her back, information about changes in direction of movement are easily conveyed to the client. It is also useful to pause briefly before sharp descents, such as off a curb or down a flight of steps.

Items 12 and 13 reflect the fact of the growing use of "signal"—both guide and hearing—dogs, and of the concurrent use of wheelchairs by some members of this population. Seating by, or in, the aisles of a meeting hall or theatre may prove essential to provide the necessary additional floor space required in these circumstances. It may also be necessary to provide someone to push the person in the wheelchair, not something that every sighted guide is comfortable doing, and to provide some kind of kennel for the dog(s).

COMMUNICATION: HARD OF HEARING AND DEAFENED PEOPLE WITH VISION IMPAIRMENTS

Item 14, as item 7, allows the coordinator to categorize quickly the communication needs of consumers who will attend the event, and to get a rough overall estimate of the resources that will be necessary. Item 15 covers assistive listening devices that are dealt with in depth in other chapters of this book; some additional attention to special modifications is warranted here.

As we pointed out at the beginning of this chapter, most people with combined hearing and sight impairments do not fall into the traditional category of deaf-blindness, but are actually hard of hearing people with the addition of visual loss. In many cases, sophisticated modern amplification devices—hearing aids and assistive listening devices (ALDs)—may suffice to provide communication access. Optimal auditory accomodation is especially essential, however, because these people rarely have the fine visual acuity necessary for speechreading, which normally sighted, hard of hearing people use to augment their residual hearing. Fortunately, especially in cases of Usher syndrome type 2, even some people with severe vision and hearing loss have substantial success through the use of ALDs, provided—and this is an important caveat—that the microphones are placely close enough to the speakers' mouths.

This last point cannot be overstated. In certain applications, such as meetings or theatrical productions, omnidirectional microphones are placed at strategic locations on a surface—floor or conference table—and they pick up multiple speakers from varying distances.

Where speechreading and residual hearing are used together, this can be quite effective. With our target population, the odds are significant that successful communication access will not be achieved this way. The use of multiple microphones is highly indicated here, preferably handheld, but lapel microphones and floor and table stands are also options. (The preference for handheld microphones stems from the ease with which these can be passed to another speaker; it is important, however, that speakers be cautioned to hold the microphones pointed toward their mouths, about 4 to 6 inches away.)

The goal is to have each speaker speak directly into an individual microphone. In fact, there is no limit to the number of microphones that can be arranged around a particular site; thus, making it possible, in some situations, to have a microphone for each speaker, or at least to place several microphones around an area, each strategically located so as to facilitate their use by all in attendance. Although a qualified audio engineer or sound technician is a good person to make these arrangements, even a novice can use easily available microphone mixing units (mixers) to "hardwire" multiple microphones into a public address or loop system amplifier, or an FM or infrared transmitter. Such microphone mixers are available from a variety of sources; Radio Shack units may cost less than $100, whereas more sophisticated units can run up to several hundred dollars. Professional performance quality wireless microphone systems (this writer happily uses a system manufactured by TOA Corporation), which are actually FM transmitters/receivers that work on a very different frequency band from the FM systems specifically intended for hard of hearing people, provide superb sound quality, and can be combined to allow a significant number of wireless microphones to be used simultaneously without the Medusa-like snarl of wires that often characterize hardwired systems. Of course, such quality and versatility comes at a high price: $600 to $900 for each FM wireless microphone, whereas some high-end conference room systems can be even costlier. (Less expensive FM wireless microphones, such as Radio Shack's, usually are not useful for our population because of the compromised sound quality.)

I have received a number of queries about setting up support groups for hearing- and sight-impaired people, and I always emphasize the need to use as many microphones as possible. Not only does this allow for optimal sound quality, it is also important because blind- or vision-impaired people can not easily "search out" a microphone that is not readily within reach. The fewer people who must share a microphone, the easier the access to that microphone. I have also found it best to have an assortment of microphones available, both omni- and unidirectional. Unidirectional microphones have the advantage of reducing the distractions of any ambient noise. They have a disadvantage in that, when the speaker lacks manual dexterity, he or she may find it difficult to point the microphone accurately toward the mouth in order to ensure the best sound quality. Omnidirectional microphones pick up sound across a broad range of directions. Therefore, I try to make omnidirectional microphones available for these individuals, while having unidirectional microphones for others: thus, achieving a balance between ease of use and reduction of background noise. Seating this kind of group at long tables also makes it easier to manage the multiple microphones, which can be conveniently set back down on the table in between turns of use.

Two of the modes listed under item 15 may give the reader pause: what is a voice shadow? and real-time captioning for people with sight impairment? Voice shadowing is actually the oldest technique. This simply involves having an SSP assist the hearing-impaired person by repeating, usually close to the listener's ear, what is being said by distant speakers. It is an especially effective tool on outdoor tours, but I have also used it with good results in certain situations where video-taped presentations are unintelligible to people with less than perfect hearing. It is also useful as an access method in impromptu circumstances where there is none of the advance notice necessary for setting up an ALD. Of course, care must be taken to reduce the effects of distracting background noise, both for the client and the worker, and also not to be a distracting source of noise for others in attendance.

Another use for voice shadowing should be mentioned. Some individuals, often because of neurologic damage concurrent to the dual sensory loss, are highly sensitive to the cadence and pitch of the voices they can understand. They may also be highly susceptible to interference from ambient noise. Essentially, this means that they learn to understand a few voices well, but poorly discriminate many other voices in running speech. An SSP who "shadows" (repeats the words of) the voices of other people at an event can, therefore, be the best way to provide communication access. In these occasions, special "noise cancelling" microphones, often on a small head-mounted boom, may be highly desirable because they come closest to truly eliminating distracting noise (obviously, these microphones can be used to excellent advantage in any voice-shadowing activity, when connected to personal or group assistive listening devices).

Real-time captioning can be useful, provided care is taken to make sure the captioning device is appropriate for the visual capability of the client. Positioning of the client and the captioned screen are crucial. Proper distance, contrast, and brightness are also key elements. Personal computers with cathode ray tubes (CRTs) rather than large screen projection or liquid crystal display (LCD) screens of most laptops (which lack both brightness and contrast) can be effective. Another option is to use one of the current generation of electronic text telephones (TT). A technician can easily install a switch to "turn off" the distracting beeping of the acoustic coupler; if the TT has a direct connect feature, simply inserting a modular plug will perform the same function.

An even more valuable modification to the TT would be installation of a large visual display (LVD, see figure 2). This is a three-inch high, 20-character long, green on blue vacuum fluorescent tube display specifically designed to make TTs legible to people with visual impairment. Because it is wired into the printer port of a TT, but is a free-standing unit, the LVD does not disconnect the built-in display; thus, both typist and viewer can be positioned to easily read the text that is being typed. Obviously, all the options we have discussed here in regard to captioning imply either small group or one-to-one typing to viewer. It should not be forgotten that some people may have severe tunnel vision but unimpaired central visual acuity; thus, they are able to use the shared viewing of captioning projected on such surfaces as movie screens.

Finally, there is that reliable workhorse, the Teletouch machine. This is a mechanical device, somewhat like an old portable manual

Figure 2.   A large video display (LVD). (Photo credit of Ultratec, Inc., Madison, Wisconsin.)

typewriter, whose output is a single mechanical Braille cell. The client and typist face each other, perhaps sitting at a slight angle, with the machine between them, and the client reads the typing, one Braille letter at a time. This invariably results in rather slow communication. If the typist is attempting to transliterate another's speech, considerable editing will be necessary, compressing the spoken message into its major points. Nevertheless, despite its somewhat antique appearance, and particularly in cases in which the person was blinded before deafened, the Teletouch can be a gateway to accessible communication.

The last line under item 15 is the ubiquitous "Other." Although standardization of access is highly desirable for the service provider, it is a simple fact that some people use home-made, school-based, or foreign-national strategies. As always, flexibility is the real key to providing communication access.

COMMUNICATION: DEAF PEOPLE WITH VISION IMPAIRMENTS

Items 16 and 17 bring us into the realm of sign language interpreters. This subset of that field—deaf-blind interpreting—is gradually beginning to come into its own, and now has its own special interest group within the national Registry of Interpreters for the Deaf.

The coordinator re-emerges as a person (or team, in many instances) of primary importance where interpreting is concerned. As we have seen, the coordinator must collect information regarding various needs of individual consumers, and skills of various support workers. If an event will be attended by many clients, and it is also likely that there will be many interpreters, the coordinator's job will be to match workers and clients of comparable skills and needs. Interpreting is a physically stressful occupation, and its practitioners have increasingly been subject to such stress-related physical conditions as carpel tunnel syndrome. This is true of those who interpret for normally sighted deaf people; tactile interpreting is considerably more fatiguing. Thus, an optimal setup will allow for rotation of interpreters about every twenty minutes to half an hour, and 5 to 15 minute breaks for both interpreters and clients should occur at least every hour, if possible.

There is the additional important goal of assessing the characteristics of the physical site where the event will take place. Good lighting is imperative. It should be strong, but efforts should be taken to avoid glare, which can be painful to individuals with such common vision impairments as cataracts and RP. If spotlights can be focused on the platform interpreters (discussed below), so much the better.

Contrast is almost as important. Interpreters should be reminded to wear dark, simple clothing so that their hands can be seen clearly

and without distraction. Similarly, simple dark backgrounds will facilitate visual perception. At conferences or meetings where the speakers
will be on a raised stage, it may be useful to tack a large black cloth (a
plain sheet or bedspread will do) to the wall behind the stage. Usually,
there will be a "platform interpreter" standing at the side of the stage.
At meetings of the American Association of the Deaf-Blind, there are
two such interpreters, one at each side. One of these interprets speech
into American Sign Language, while the other interprets into an
English-based sign system. It is best that they, too, wear simple, dark
clothing, and stand in front of a background that enhances their visual
presence.

Items 16 and 17 on our sample application form make it clear
that more interpreters are needed for our population than just the platform interpreters. If the speakers will use sign language, a lead voice
interpreter will also be needed, preferably using a microphone hooked
into both a public address system and some large area assistive listening device. The platform interpreter will also copy the signs of ques-
tioners/speakers from the audience. This is an important part of his or
her role because there will be interpreters for individual clients who
will receive their information from either the platform or lead voice
interpreter.

Before proceeding through our list of the different interpreting
options, a unique point must be emphasized: interpretation for visually
impaired people, most especially for those who are fully blind, is not
only a process of conveying linguistic data. Increasingly, interpreters
within this specialty rise to the challenge of providing access by communicating important visual data about the circumstances and people
of the event. Ideally, interpreters meet their clients before the proceedings begin, orient them to what might be significant in the details of
the location (who, having seen them, has not experienced an emotional
impact from the visual aspects of a large and beautiful theatre, stadi-
um, or even banquet hall?). During the actual proceedings, some
thought should also be given to communicating significant visual
details about speakers, actors, and action.

I have a strong memory of sitting in the rear of a huge banquet
hall while tributes were being spoken and signed to the outgoing president of the American Association of the Deaf-Blind. At one point, the
voice interpreter, who was using a microphone that was, very appro-
priately, connected to both PA and FM systems, added to her com-
mentary the observation that the guest of honor was crying in response to the numerous honors being bestowed upon him. This information, which was clearly available to normally sighted members of
the audience, thus became available to the hundreds who were
involved in the action through the hands of their interpreters, and to

me, seated in the rear, and hooked in with my FM reciever. Our experience was enormously enriched, far more than if we had been given only the words of the gentleman.

The next two options under item 16 are "close visual interpreter," both in small groups and one-to-one. It is common for groups of 4 to 6 individuals with similar visual capacities to be able to share an interpreter, sitting in an arc or short row before her or him. Clients and interpreter should arrange themselves at an optimum distance, balancing for visual field, acuity, and the space constraints of the room. Also, as always, seek to locate the group where lighting is clear, but with minimal glare and distraction. A related and important access function for interpreters is in the integration of the deaf/visually impaired person into "mixed" discussion groups, which also contain hearing and deaf people (Alcoholics Anonymous provides such mixed groups in some locales). Even if the other members of the group sign, the visual impairment may make adjusting to—and locating—other signers difficult or impossible for the client. In that case, as well as for nonsigning group members, the interpreter will sign all linguistic data, and should identify who is talking at the moment by pointing toward and/or naming the speaker.

Others, usually with more significant impairment of visual acuity, need to sit close to and directly in front of their interpreters. Some may not want to share interpreters because they are selective about the kind of language they prefer to use: if for example, they prefer a significant amount of English vocabulary and grammar, or they prefer presentation of sign language and fingerspelling to be somewhat slower than others desire.

The production and comprehension of tactile interpretation is usually a slower process than visual sign language, although native ASL users are usually more proficient with this skill than are others. Note that we have included, in item 17, three language options that can be either visual or tactile sign language, as well as two languages for fingerspelling. The American "one handed" fingerspelling technique is still used by some deaf-blind people. Some of these people seem to be able to comprehend extremely fast fingerspelling, and can also receive it in more than one position—facing their interpreter, and when sitting or walking side-by-side with their interpreter are two examples. Increasingly, however, those who have lost hearing and vision find it significantly easier to learn some sign language method. The English "two handed" fingerspelling technique is probably somewhat less taxing for interpreters, but still slower than sign languages. It is included here because Canadians and Britons attend functions in the United States, and have learned this method.

Note that in all tactile interpreting, people tend to have a domi-

nant hand for receiving, and if possible, should be matched with interpreters whose own dominant handedness is appropriate. Indeed, it is
interesting to note that most people learn to use only one hand to
receive sign tactile language, although their interpreters may be
observed using both hands to produce that information. Clearly, there is
a great deal of extrapolation being done by the deaf-blind person who is
comprehending sighted languages soley through the more spatially limited tactile sense—an interesting area for cognitive scientists researching
neural modeling! In any event, this skill allows most, but not all, inter-
preter/deaf-blind dyads to be seated next to each other. A table for resting their elbows, whether they are seated facing or beside each other, is
a good idea. I have seen a few people who have learned to work together
as a team and found small pillows or pads useful for cushioning.

Tracking, under item 16, is a way for people with tunnel vision
to use their hands to assist in visually perceiving sign language or fingerspelling. The receiving person uses one of his or her hands to lightly
hold the other's elbow or arm. This keeps him or her from losing sight
of the other's sign language presentation. Obviously again, people
who use tracking will require their own interpreter. Moreover, when
interpreting for individuals with tunnel vision, such as those with
Usher syndrome, interpreters should always be alert to modifying
their normal use of signing space.

Again, it is necessary to check with individual clients to determine the extent of the remaining visual field, and then to tailor the
signing space to that. Because, as a visual language, ASL places specific
grammatic and syntactic meaning in this space, the interpreter may
need to make further modifications, replacing a sign that is more easily
seen and accurately understood in this space for another sign, perhaps
more expressive, but inappropriate to the visual capability. This is precisely analogous to modulating one's voice and using easily intelligible words in communication with hard of hearing people.

Item 16 also has a line for "Tellatouch or other typing." It is helpful to list these options under both "hard of hearing" and "use interpreter" categories, because some people may look under only one category and miss this type of service. Item 17 lists two other options.
Tadoma is an infrequently used option, an extraordinary tactile fingerspelling technique learned by few, and taught today to even fewer. The
receiving person places his or her thumb against the other person's
lips, and outstretched remaining fingertips of the same hand lightly
touch the speaker's throat. Skilled individuals can be surprisingly successful in discriminating speech. Again, it is obvious that Tadoma users
must work one-to-one with an interpreter. Oral interpreting, discussed
elsewhere in this volume, may be useful given the same kinds of constraints and considerations to optimal visibility that have already been

noted. "Other" may prove an important line in item 17. People who have been deaf and visually impaired from early in life are perhaps even more likely than those who are deaf, but normally sighted, to have acquired a home-made language/communication system, perhaps employing some of ASL, English, Spanish or another language, home sign, and gesture. A few interpreters have acquired skill in meeting the remarkable challenge of communicating with this kind of client. The final three items (18–20) should be self-explanatory; even where neither ADA-mandated nor concerned with access per se, experience has proved their usefulness in ensuring smoothly run functions.

TELECOMMUNICATIONS ACCESS

The ADA also recognizes the rights of all citizens to have full access to our national telephone networks. Third-party relay services are mandated in all states. Free provision of adaptive telephones—amplified, text telephones, and Braille output phones—is not required, however. Nevertheless, recognizing that the additional expense of this type of equipment may place an undue burden on hearing-impaired people, some states have used the excise taxes that pay for the relay services to also provide this kind of equipment, whereas in other states, private and public agencies have been tapped as funding sources.

For hard of hearing people with vision loss, the same amplification and frequency enhancement techniques developed for normally sighted people will serve well. However, it should not be forgotten that vision is necessary for dialing and for using telephone directories. In regard to the latter, an individual who has been certified by a physician or rehabilitation professional as legally blind or seriously visually impaired can, on application to a local telephone company and long distance carrier, receive free directory assistance privileges. As for dialing, some amplified telephones have large key numberpads. There are also add-on large keys and overlays to convert numberpads and rotary dials to larger formats, but these may prove inferior to the telephones with built-in large keys. Designing and marketing easy-to-read phones should be encouraged, and those programs that have been providing only free telephone amplifiers or amplified handsets should consider distribution of such entire telephones, lest they solve one problem but leave another unresolved.

Text telephones, for deaf and profoundly hearing-impaired people, are precisely and entirely visual solutions to the problem of access. Thus, they have substantial potential for being troublesome to our population. Perhaps it is somewhat surprising that they are not more of a problem then they actually are. The standard keyboards used on most models are not easy to read, and developing high contrast or oversized keys would be a good idea—there are some special computer keyboards of this type for both those with vision impairments and those with dexterity impairments, but I know of none for TTs. However, many people simply learn to touchtype, and for fully blind people there are few practical alternatives to this strategy.

The green on blue displays on most TTs (actually called vacuum fluorescent tubes) generally are fairly bright, and many people, even those with some impairment in visual acuity, can read these. Unfortunately, some TTs, such as laptop computers, employ black on gray display screens, which should be avoided where visually impaired people are involved. On the other hand, as we noted above, there is an LVD on the market, with a three-inch-high, bright-red display that can be hooked into the printer port of specially modified TTs (the modification is easily done by the manufacturer or any technician familiar with repairing the equipment). These LVDs have a full, 20-character, three-inch-high display, and they cost less than $300—thus, they are a simple and affordable solution to the problem of visual access to TTs.

More complex solutions to this problem are available when the consumer has a personal computer. Several companies now market computer modems that are TT-compatible, allowing the computer user to communicate directly with other hearing-impaired people who have TTs. Both software and hardware solutions are now available to enlarge the size of output displayed on the computer monitor. Some of these software programs are: *See Tec Magic, Zoom Text,* and *LP DOS.* They can be obtained from a number of services offering assistance to visually impaired computer users, who can easily be located through your state rehabilitation agency for blind and visually impaired people.

There are also several hardware options. Several companies now produce 20- and 26-inch computer monitors, that automatically produce larger text with good control over contrast and brightness. One company, TeleSensory Systems Incorporated (TSI) in Palo Alto, California, makes a Vista plug-in board to allow computer users to expand the size of text displayed on their monitors greatly. That company, and others, also manufactures closed-circuit TV reading systems, that can be interfaced with computers and allow for as much as magnification to the twentieth power of text on the monitor, providing for a sophisticated text telephone.

Using a computer is a sophisticated and effective, but expensive, solution to the problems of telecommunication access. At least as expensive are the Braille output devices used as telephones for deafblind people. One of these, the Infotouch Embossed Telebraille from Enabling Technologies (see figure 3), costs about $4,000; another, the TeleBraille III from TeleSensory (see figure 4), is about $6,000.

Figure 3.   Infotouch Embossed Telebraille. (Photo courtesy of Enabling Technologies Company.)

Actually, both of these devices are a merger of one of the standard TTs on the market, with one of two Braille output devices developed for use with computers.

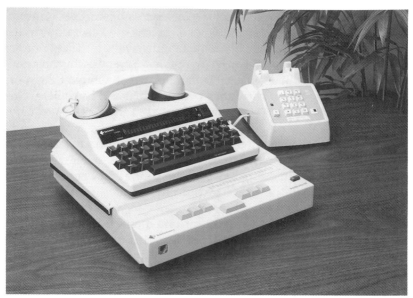

Figure 4.   TeleBraille III. (Photo courtesy of TeleSensory.)

Infotouch uses Ultratec's Superprint Model 100 D, which has a full complement of modern TT features: standard keyboard, direct connect, built in answering machine, and so forth. This is connected via cables to Enabling Technologies' own Romeo Braille Embosser, which produces fully readable Braille text on paper, at a maximum speed of 20 characters per second (although, of course, when used as a telebraille, the Romeo will only print as fast as the other party types). The deaf-blind consumer must wait until at least a full line of text has been printed, at which point the machine will scroll to the next line and the deaf-blind person can only then use his or her fingers to touch and thus read the text. A considerable time lag is created by this process; rather than being able to read the message as it is being typed, as in standard TT communication, the communication here entails having the deaf-blind person always at least a line behind the person who is typing. This machine also takes up a lot of space, because the Romeo is the size of a small suitcase, and it must sit separately from the TT on a table. Nevertheless, Infotouch is quite reliable and does its work in a straightforward manner, allowing real telephone communication for deaf-blind users, who also have the advantage of automatically getting paper copies of all of the conversation.

TeleBraille III may cost half again as much as its competitor, but it takes up less than half the table space. The TT, in this case the Supercom, rests on top of its Braille module; thus the smaller footprint. The Braille module for this device is a 20-cell "refreshable" display. There are actual pins that emerge through the six holes that make up each cell, denoting the six possible points of a Braille character. There is no paper copy, although a printer could be attached to the TT's printer port. The user reads the text directly off the pins of the display cells. It is, therefore, almost possible for the user to communicate in "real time," reading each Braille cell as it emerges at the same speed it is being typed. At least, in theory that's true. The reality is that only the very fastest Braille reader can even approach the speed of an ordinary typist. TeleBraille III has a unique feature, the "unlink" command, which allows the user to save text that is being typed, almost always too quickly to be easily read, into a memory buffer. Then the deaf-blind person can scroll through the message in the buffer at his or her own pace.

The result of this procedure is the same kind of lag that we saw in Infotouch. At least with TeleBraille III it *is* possible to avoid this lag, if, and only if, the person typing to the deaf-blind person types slowly, carefully tailoring typing speed to reading speed. What gets really complicated is communication with a telebraille user (using telebraille in the generic sense, denoting either of these devices) via a third party relay service. Here we have one person speaking, another person typing, and finally the deaf-blind person reading the Braille output. Not only can lag become an extreme problem here, it is simply not uncommon for businesses who recieve calls from deaf-blind telebraille/relay service users to hang up in exasperation. Imagine, also, the complex challenge for a deaf-blind person seeking to use the relay service to leave a message on someone's voice mail system! An additional problem arises when the relay operators (communication assistants [CAs]) use the function keys on their terminals to automatically produce some of the standard messages that are part of the relay service. These automatic messages are generated quickly. They are difficult to read visually and because of their speed, impossible to handle tactually.

Given that universal third-party relay service in the United States is still in its infancy, and widespread distribution of telebrailles is barely out of the toddler stage, it is not surprising that there are problems interfacing the two. The state of Washington, in response to feedback from Seattle's large and politically active deaf-blind community, has begun to institute certain measures designed to improve the use of telebrailles for accessing the telephone network. This state is providing telebrailles free-of-charge, using the fund created by the excise tax that also supports the state-wide relay service. Individual training is provided to each of these consumers, who also have the right to choose either Infotouch or TeleBraille III. A separate telephone access number for telebraille users was instituted (with additional separate numbers for voice and TT users) in the summer of 1993. When answering this line, CAs are trained to type manually (no function keys), slowly, and to fill time gaps with messages to the voice user, such as "Relay here, please hold." Deaf-blind users are also being encouraged to participate in the education process themselves, helping to inform hearing and deaf people about the benefits and limitations of the unique methodologies employed for their telephone access. Some of this education occurs by having two representatives on the advisory committee for the state Telecommunication Access Service: one fully deaf and blind, the other partially sighted and hearing. Clearly, further thought, creativity and experimentation with all those concerned is in order, so that even better access can be achieved.

SIGNALING DEVICES

Access to the national telephone network certainly must begin at home. Sensory-impaired people must have some reliable way of knowing when there are incoming calls on their equipment. The loud, tone-selectable ringers designed for hard of hearing people may work well for some people with additional visual impairments. A household equipped with a strategically placed series of flashing lights may be a haven for deaf people with impaired eyesight who feel comfortable knowing that they will be alerted when someone is on the telephone or at the door.

Most of those with whom we have been concerned will, no doubt, feel even more comfortable if they and their homes are outfitted with vibrotactile alerting systems. Three such systems currently are on the market. The first system of this type was developed over ten years ago at the Helen Keller National Center for Deaf-Blind Youths and Adults, at Sands Point, Long Island. This is the Tactile Communicator (TC), now manufactured by the Sonic Alert Company, and also called the Vibrating Signal System. There are at least three components to this system.

1. *A Sensor.* This is a small box that can be wired directly in contact with the signal source; the telephone, doorbell, and smoke detector are the three primary sources.
2. *A Transmitter.* Larger than a sensor, this is actually a base station. As many as eight sensors can be set up to transmit their signals (via FM) to the transmitter, which will then rebroadcast the information, along different FM frequencies, to the receivers.
3. *A Receiver.* About the size of a cigarette pack, these receivers are worn somewhere on the body of the user (in a pocket, or hip or belt pouch, for example). The receiver itself vibrates, and the user must learn to recognize different vibration patterns—somewhat like Morse code—that indicate which of the signal sources is active. Thus, three short bursts might mean the telephone, two short and one long the front door, two long and one short the back door, and three long bursts might represent the smoke detector.

The Tactile Communicator has a steep price tag, about $520 for a system with a sensor for the telephone only, but it is an effective way of providing this initial level of telecommunications access. Adding an additional sensor for the doorbell or smoke detector will raise the price $40 to $90, but this system will now provide accessibility in public accomodations, such as hotels and motels, as required by the ADA.

Two other recently developed systems function similarly to the TC. Quest Electronics manufactures the Silent Page, and Silent Call Corporation makes the Silent Call. Both of these systems differ from the TC in eliminating the base station transmitter; their sensors transmit directly to the receivers. Each of these companies actually makes two receiver models. One model vibrates to alert the user, but then uses light emitting devices (LEDs) to specify the signal source. The other model is a fully tactile unit, with differing vibration patterns. The Silent Page receiver is smaller and wrist-worn; the Silent and

Vibra-Call units are close to the size of the TC, although the manufacturer has been promising for years to further miniaturize its products. However, whereas the TC and Vibra-Call have rechargeable battery options, the Silent Page requires replaceable batteries that are either difficult to install or difficult to locate in stores. Moreover, the Silent Page tries to make itself easier to set up by using a special selectable sound monitor feature, thus not necessitating the direct connect wiring of its two competitors. Unfortunately, it seems that this method of set-up is less reliable than the direct connect method, and the Silent Page has a reputation of being prone to false signals. All three systems cost about the same. Trade-offs can be made in selecting any of these systems, and all are sleekly tooled, sophisticated solutions to the problem of providing ADA-mandated access to the telephone networks and public accomodations.

THE FUTURE

Provision of accessibility for people with combined sight and hearing disabilities has clearly gone forward in leaps and bounds in recent years. With passage of the ADA, this progress should continue. Indeed, the Rehabilitation Services Administration has been placing "deaf-blindness" higher on its list of priorities, and other institutions, both academic and direct service, have been attending to the many dimensions of this chronically underserved, but clearly numerous, population. Nor is the engineering community resting on its laurels. There have been efforts to develop several different local personal computer bulletin boards specifically for these users, to develop useable Braille output closed-caption television decoders, and to refine and market a mechanical fingerspelling hand to allow for direct communication between deaf-blind people and those who lack manual communication skills.

However, along with these positive indications, are new hurdles of which to be aware. The movement toward computers that use voice recognition for input, and synthesized speech for output, could pose new barriers. Even within the very field designed to serve our special consumers, there can be oversights stemming from an overly simplistic view of these consumers. The TeleBraille II, predecessor to the model we discussed above, employed a TT with a difficult-to-see LCD display on the assumption that if one uses Braille, visibility is of no importance. On the contrary, there are many individuals with fluctuating vision, who may use Braille during the afternoon, but be able to see a well-contrasted visual display at night—and thereby increase their reading speed significantly. Unfortunately, there are still some

workers in the many fields of deafness who engage in the old wars—
oral versus manual, ASL versus English.

Only when we keep our thinking fresh, flexible, and compassion-
ate, will we be able to assure that all, not some, have access to the full
spectrum of beauty and power of life in America.

RESOURCES

Helen Keller National Center for Deaf-Blind Youths and Adults
111 Middleneck Road
Sands Point, NY 11050
(516) 944-8900 V/TTY
Comprehensive services throughout the USA. Has ten regional offices, a Technical
Assistance Project, a Parents Network, and publishes numerous resources.

American Foundation for the Blind
15 West 16th Street
New York, NY 10011
(212) 620-2000 V/ 620-2150 TTY
Source for the Telatouch machine and other assistive devices. Has a Deaf-Blind Project
to disseminate information regarding communication and mobility skills.

Gallaudet University
Hearing/Vision Impaired Program
Kendall Green
800 Florida Avenue N.E.
Washington, D.C. 20002
(202) 251-5044 V/TTY
Publishes information regarding communication with deaf-blind people and adjust-
ment to deaf-blindness, particularly Usher Syndrome.

Registry of Interpreters for the Deaf: Special Interest Group in Deaf Blindness
9703 47th Place
College Park, MD 20740-1470
(301) 314-7684
Resources about sign language interpreting.

American Association of the Deaf-Blind
814 Thayer Avenue
Silver Spring, MD 20910
(301)588-6545 V/ 523-1265 TTY
Consumer organization, sponsors yearly national conferences and published periodical
"The Deaf-Blind American."

TeleSensory
455 North Bernardo Avenue
PO Box 7455
Mountain View, CA 94039-7455
1-800-227-8418
Manufacturer of the TeleBraille III, Closed Circuit Television Reading Systems,
Navigator Computer Braille Output, Vista Large Print Computer System, and
other devices.

Enabling Technology Co.
31402 S.E. Jay Street
Stuart, FL 34997
(407) 283-4817
Produces the Infotouch Embossed Telebraille and the Romeo Computer Braille Printer.

Ultratec
450 Science Drive
Madison WI 53711
(608) 238-5400 V/TTY
Manufactures TTY's with Large Visual Display.

LS & S Group, Inc.
PO Box 673
Northbrook, IL 60065
1-800-468-4789
*Source for large print computer screen enhancement software, as well as many other
    assistive devices.*

The Technological Future

Chapter • 14

Communications Technology and Assistive Hearing Technology
Future Trends

Harry Levitt

BACKGROUND

Communications technology and assistive hearing technology[1] have developed symbiotically over the years. A. G. Bell, in his memoirs, reports on how his attempt to develop a visual display of speech for his deaf wife led to the development of a technique for converting acoustic vibrations in air to mechanical vibrations of a solid object. This discovery, in turn, led to the development of a practical telephone. In this early example of symbiosis, communications technology benefitted substantially from research in assistive hearing technology. The reverse situation, however, is far more common in that assistive hearing technology is usually the beneficiary from advances in communications technology.

The history of the hearing aid illustrates the nature of this relationship rather well. The need to amplify sound for telephone and radio applications resulted in the development of electronic ampli-

Preparation of this chapter was supported by Grant #H133E80015 from the National Institute on Disability and Rehabilitation Research, U.S. Department of Education, to the Lexington Center for a Rehabilitation Engineering Research Center (RERC) on Hearing Enhancement and Assistive Devices.
[1]Assistive hearing technology is that branch of technology concerned with the development of devices for people who are deaf or hard of hearing.

fiers. Among the first applications of this technology outside of radio and telephone engineering was that of an electronic hearing aid and an electronic instrument for measuring hearing impairment (the audiometer).

The development of miniature vacuum tubes resulted not only in smaller, more economical electronic systems for the rapidly growing telephone and radio industries, but also in the development of personal, wearable hearing aids. Similarly, one of the first commercial applications of the transistor outside of the communications industry was in the development of miniature hearing aids. Hearing instruments small enough to fit behind the ear were introduced at about the same time as compact transistor radios. The ongoing trend toward the miniaturization of electronic circuits resulted in the development of tiny electronic chips that are now used for a wide range of industrial and commercial applications, including the development of hearing aids small enough to fit in the ear canal.

The electronic chips developed for hearing aids differ from those used for other audio applications in that, because of the limitations on battery size, these chips must operate at low voltages and draw very little power. The development of hearing-aid circuits has thus evolved into a specialized area of chip technology that has been driven largely by consumer demand for hearing aids of smaller and smaller size.

Another invention that has found an unexpectedly wide range of uses outside its intended application is the digital computer. This invention was initially conceived as a device for performing computations efficiently at high speeds and with great precision. It was soon realized, however, that digital computers could also be used to sort, store, and retrieve vast quantities of information, perform logical operations, monitor, analyze, and synthesize a wide range of signals, including audio and video signals, and even to process natural language. Given this tremendous range of capabilities, the applications of computer technology have mushroomed, affecting almost every aspect of our lives.

A concomitant development that is perhaps equally dramatic is the remarkable progress made in both miniaturizing and reducing the cost of the components used in modern computers. As a consequence, the digital chips and other components initially developed for computers are now being used in an extraordinarily wide range of applications. The telephone industry is in the midst of a technological revolution in which traditional analog methods of processing, switching, and transmitting voice signals are being replaced with advanced digital techniques. This transition will allow a wide range of new capabilities to be implemented in the telephone system. The technology developed for this purpose is also being transported to the hearing aid industry, and several manufacturers have recently introduced digital techniques in their hearing aids.

The second most widely used form of assistive hearing technology, after the hearing aid, is the text telephone. As in the case of the hearing aid, the development of the text telephone was heavily dependent on both consumer demand and related advances in communications technology.

The text telephone evolved from the teletypewriter (TTY). The TTY has been in use in commerce and industry for well over half a century. Although the TTY could also be used by deaf people as an alternative to the telephone, the extent of this usage was severely limited, at least initially, because of the relatively high cost of the equipment and the need for both parties in the communication link to have a TTY. In 1964, a deaf physicist invented a device for transmitting text using a conventional telephone. Letters and other symbols were converted to acoustic signals that were then picked up by a conventional telephone handset using an acoustic coupler. These signals were then transmitted over the telephone network, in the same way as other audio signals, to a telephone at the receiving end that converted the acoustic signals back to text and then displayed the text visually. This device, known as a TDD (telephone device for the deaf) was much less expensive than a TTY and, as a consequence, was affordable to large numbers of deaf people.

A concurrent development was the growing use of telephone lines for transmitting information to and from computers. Although the same basic method of converting alphanumeric information to audio form for telephone transmission is used, a different audio code is usually employed for computer communications. Computers use the ASCII code, which is much faster (but less robust) than the older Baudot code used by TTYs and early TDDs. Because the two codes are not compatible and the faster ASCII code has become the standard for modern electronic information networks, owners of older TDDs are unable to access these information networks. Several states, in their distribution of TDDs, have included both ASCII and Baudot in the instruments provided.

A positive benefit of the rapidly growing use of computers and communication systems is that devices for electronic communication are being mass produced in increasing numbers with concomitant reductions in their cost. As a consequence, the cost of TDDs has dropped significantly, thereby making these devices affordable to even larger numbers of deaf people. The most dramatic reductions in cost occurred during the 1970s, and the size of the TDD network (as measured by number of subscribers) showed substantial growth during this period.

An additional benefit of the rapid growth of electronic communication is that many features of computer communication, such as message storage, automation of tedious operations (e.g., repeat dialing), and word processing (to a limited extent) are being incorporated into modern TDDs without adding significantly to their cost. Because of the growing range of functions of modern TDDs, these devices are now being referred to as *text telephones*.

Technological innovation is not always advantageous to deaf and hard of hearing people. The invention of the telephone was of great benefit to hearing people, but was of little practical value to the deaf community until the development of inexpensive text telephones. Another invention that has had a checkered history with deaf and hard of hearing people is that of the motion picture.

The earliest motion pictures did not have sound and were of equal benefit to deaf and hearing people. The transition from silent movies to talkies, however, was a loss for the deaf community in that the heavy dependence on visual gestures was no longer used with talking pictures. (A unique art form known as "emoting" came and went with silent movies.) Similarly, the captions used with silent motion pictures also disappeared, although a new form of captioning specifically aimed at a deaf audience was re-introduced at a later date.

The demand for captioned motion pictures, including foreign films with subtitles, is small relative to that for motion pictures in general. The cost of captioning is also relatively high and, as a consequence, few motion pictures are produced with captions or have had captions added to them. Television captioning is not as expensive, but this form of captioning was not widely used until consumer demand, coupled with both corporate and governmental support, led to the development of a practical means for captioning television programs. The system currently in use transmits the captions separately from the video picture so that captioning is provided to only those people who want captions and who have the necessary decoder.

The use of television captioning has grown rapidly with the availability of inexpensive decoders for captioning. The cost of producing the captions, however, is still relatively expensive and the service is subsidized to a large extent. The now widespread use of television captioning is a new development resulting from a combination of consumer demand, market economics, and both government and corporate subsidies. The technology for television captioning could have been developed much sooner had these forces been in place earlier.

The history of the hearing aid, text telephone, and television captioning illustrates two important principles: the mutual interdependence of different branches of technology and the importance of consumer demand. Whereas invention may give birth to a new technology, the nurturing and growth of this technology is critically dependent on consumer demand. The relatively slow early development of TDDs, for example, was not a result of technological limitations but rather one of limited consumer demand determined largely by the high cost of early TDDs. Similarly, captioning of movies and television was limited initially by relatively high costs and low demand. As a result of growing consumer demand, which in turn generated both governmental and corporate support, a means was found for harnessing this service to regular television programming. In the case of hearing aids, the trend toward developing smaller and smaller instruments was not driven by audiological considerations, but rather by consumer demand for instruments that would be barely visible and hence cosmetically acceptable.

Finally, it should be noted that the underlying reason that assistive hearing technology has benefitted more from advances in communications and computer technology than vice versa is that there is a much greater demand for the latter technologies by the general population. Stimulated by this demand, which is substantial, a major research and development effort has been directed toward advancing communication and computer technologies. The results, thus far, have been dramatic with significant implications for the future development of assistive hearing technology. In order to take advantage of these developments, however, it is important to find a practical means of harnessing new technologies for the benefit of deaf and hard of hearing people.

CURRENT DEVELOPMENTS IN COMMUNICATIONS TECHNOLOGY

Both the telephone and television industries are undergoing profound change. These changes are bound to have a significant impact on assistive hearing technology, and it is important to take note of this changing landscape.

Perhaps the most important change taking place is the merging of computer technology and communications technology. Whereas digital computers have been used for controlling telephone switching circuits for some time, the current trend is toward computerization or digitization of all aspects of the telephone network. That is, digital techniques will not only be used for switching and controlling the signals being transmitted in the telephone system, but the signals themselves will also be converted to digital form and then processed and transmitted using digital techniques. This conversion has already been initiated by most telephone companies.

The advantages of an all-digital telephone system are many, including improved sound quality for speech signals and greater speed and efficiency of transmission for computer communications, as well as a range of new options. Perhaps the most important advantage is that the telephone system will become a general-purpose information transmission system in which information of all kinds (text, speech, music, video, data) can be transmitted efficiently over the same network.

Telephone companies world wide are gradually moving toward an all-purpose digital communications system known as the Integrated Services Digital Network (ISDN). This system will handle digital information of all kinds. In this context, information should be viewed as a quantity that has many applications and that can be purchased in small or large amounts in much the same way that electricity is purchased. The unit of information is the *bit*. A text telephone for example, would require relatively few bits per second for its operation, whereas the transmission of video pictures over the same telephone line would require several thousand times as many bits per second. If the cost of the telephone service is roughly proportional to the information transmitted (per unit time), then the cost of using a text telephone should be a small fraction of the cost of using a video telephone.

The information rate for transmitting speech over a digital telephone line is greater than that for text, but less than that for video signals. On the other hand, transmission of pictures at a slow rate, such as using a FAX machine for picture transmission, involves a rate of information transmission comparable to that of speech. If improved sound quality is desired, such as transmitting music with high fidelity, an information rate several times that required for current telephone-quality speech is needed. The above examples should serve as a rough guide of the relative cost, in terms of information rates, of the various signals that are likely to be used in a general-purpose telephone system.

The television industry is also undergoing a major technological revolution. Not only will the quality of television pictures be enhanced considerably by the likely introduction of high definition television (HDTV) in the near future, but the growing use of digital technology in processing video signals will result in substantial improvements in both the control and transmission of television pictures. Many television and VCR systems now provide text and pictographs that appear on the video screen in order to help the user adjust the set. Similarly, many telephones now have small video screens that provide information to the user, such as the number being dialed, or the number of the caller or, in the case of pay telephones, helpful information on how to use the telephone. It is likely that interactive communications of this type will grow as the complexity of communication systems increases.

Perhaps the most dramatic change taking place in the television industry is the rapid growth of cable networks for television transmission. Television was limited initially to the broadcast media. In this form of information transmission, signals from a few sources (e.g., from regional or national radio and television stations) are transmitted to a large number of receivers within the transmission range of the broadcast station. This form of signal transmission does not require the use of wires to each receiver and, as a consequence, is an extremely convenient form of information transmission. The convenience and popularity of broadcast systems is evident from the large number of car radios and portable receivers (radio/TV) in everyday use. A fundamental limitation of broadcast systems, however, is that information is transmitted in one direction only, from transmitter to receiver, with large numbers of people receiving exactly the same information at the same time. One-way information transfer of this type is also characteristic of movies and newspapers. These forms of communication have been aptly termed mass media because of the large number of people receiving information by these means.

With cable television, it is possible for the person at the receiving end to send information back, thereby setting up a two-way communication channel. This is technically possible because the signals reaching each television receiver are conveyed by cable, and the same cable can be used to convey the response to the received information. A simple form of two-way communication using cable television is already in existence with the newly introduced shopping television networks in which viewers of these channels use the telephone to call in their orders.

Two-way television communication is being used increasingly for educational applications in which a lecturer is at one location and students are in several geographically separate classrooms. The lecture is televised and transmitted to all students simultaneously. Students can interact with the lecturer either by calling in their questions by telephone or, in some cases, by means of television transmission in the reverse direction from one or more of the classrooms.

Another form of interaction using cable television is that of an interactive information channel in which the viewer can select which information is to be summarized and which information can be reported in greater detail. The system is much like an electronic newspaper. The headlines appear first, then, using a remote control similar to that for switching television channels, detailed information on specific headlines can be called up on the screen. This type of interaction is possible if many channels of text are transmitted simultaneously within a single television channel. As noted earlier, the information capacity required for a text channel is very small compared to that of a video channel; hence, many text channels can be transmitted simultaneously in the space allocated for a single television channel.

The above form of interactive information transmission differs from computer information services using telephone lines, in that the computer system transmits the information as requested rather than transmitting many channels of information simultaneously. In the former case, a two-way communication network is needed. In the latter case, one-way information transmission is used with the recipient simply switching among available channels. The computer-based information service has a considerably larger database to draw upon, but for certain applications, such as a news broadcast, a one-way information transmission system can be both convenient and effective.

At present, the telephone network has the valuable advantage of an immense switching system that can connect any telephone (or related device, such as a FAX machine or computer modem) to any other telephone in the world that is part of the international telephone network. The growing television cable network does not have this tremendous switching capability, but some degree of switching is nevertheless technically feasible and could be of great benefit to subscribers.

It should be noted that while cable television is able to make inroads on some of the traditional uses of the telephone, it is also possible for modern telephones, especially those that have switched to digital methods of signal transmission, to provide services previously limited to cable networks. Specifically, video telephones are now commercially feasible, and several telephone companies have recently introduced instruments of this type.

A video telephone known as the Picturephone® was first introduced by ATT about thirty years ago, but it was not a commercial success because of the high costs involved. This was largely because of the high cost of the switching systems and the need for improved wiring between the video telephone and the local telephone exchange. Given the rapid growth of both the television cable industry and computer information networks, the cost of installing and operating a video telephone system has dropped appreciably. It now seems likely that video telephones may become a reality, with the signals being transmitted by either cable or telephone wires, or (most likely) by some combination of the two.

The rapid growth of telecommunication networks is not limited to systems that are connected by wire or cable. The use of radio transmission for two-way communication is also growing rapidly. This has been made possible by the use of communication satellites that allow for radio transmission at very high carrier frequencies. Previously, signals transmitted at these very high frequencies were limited to line-of-sight transmissions and were of limited value for transmission over long distances on the earth's surface. However, with the use of a communications satellite to reflect these signals back to earth, it is possible to use these high-frequency carrier signals for point-to-point signal transmissions. Several new forms of telecommunication have resulted from this important technological advance. These include cellular telephones, paging services, location finders, and various forms of signal monitoring.

The cellular telephone is perhaps the most well-known product of this new technology. An important advantage of the cellular telephone is that it is wireless and can be used from a wide range of locations depending on the calling area covered by the telephone company. Thus, it is possible to use a cellular telephone while travelling in a car, bus, train, ship, or plane, while relaxing at the beach, hiking in the woods, or from any remote location within the calling area. The use of cellular telephones has grown dramatically over the past few years. Furthermore, long-term projections predict substantial future growth; cellular telephones may one day rival conventional telephones in terms of relative use.

Cellular telephones should not be confused with another form of wireless telephone, the cordless telephone, which uses low-power radio transmission with a carrier frequency close to the broadcast frequency range. These cordless telephones can operate only within a short distance of the base station. They are useful within a home or office, but they cannot be used if the telephone is more than one or two hundred yards away from its base transmitter. Cordless telephones and cellular telephones use different forms of wireless transmission that are not compatible with each other. It is possible for a person using a cordless telephone to speak to someone using a cellular telephone, but the signals must be routed through the conventional telephone network. It is not possible for a cordless telephone to transmit its radio signals directly to a cellular telephone.

Other communication systems that use a technology similar to that of cellular telephones include beepers and message services that can reach the user wherever he or she is located at any given time. This service is particularly valuable for doctors and others who need to be contacted immediately in the case of an emergency. A useful feature of these services is that messages can be conveyed to the user in the form of short segments of text. Other possible applications of this technology include location finders that will provide relatively accurate information about the sender's geographic location. A variation of this technology is to monitor the location of a miniature transmitter that has been mounted on a moving object for the purpose of tracking the object's movements.

An emerging branch of communications technology is that of human–machine communication. Rapid strides have been made in recent years in developing machine-to-speech and speech-to-machine communication systems. Computer methods of processing, storing, and synthesizing speech are now well developed, including the development of special-purpose computer chips for these applications. It is now possible to generate intelligible speech from text using these systems, although the quality of this speech has an artificial, machine-like quality. The reverse process, that of converting speech to text, is not as well developed, although this can also be done under limited conditions.

ASSISTIVE HEARING TECHNOLOGY: FUTURE TRENDS

The changes taking place in the telephone and television industries have already begun to influence assistive hearing technology in dramatic new ways. Digital techniques are now widely used in consumer audio products (e.g., the compact disc), and this technology is beginning to be used in hearing aids. A precursor of things to come was the all-digital hearing aid, known as the Phoenix, which was introduced several years ago. Although this instrument was not a commercial success, it paved the way for introducing digital technology to the hearing-aid field. Many modern hearing aids now use digital techniques to a limited extent. These instruments typically employ a digital unit for controlling the electronic components (amplifiers, filters) of the hearing aid. More advanced digital hearing aids in which the audio signal itself is digitized and then processed in digital form are currently being developed by several companies.

Digitally controlled hearing aids offer important advantages over conventional hearing aids. These include programmability, memory, and automatic frequency-response adjustment. In a programmable hearing aid, the electroacoustic characteristics of the instrument can be adjusted (programmed) electronically. This is usually done by connecting the hearing aid to a programming unit (a small computer is often used for this purpose) that then adjusts the hearing aid so as to provide an appropriate fit for the user. This method of fitting a hearing aid is both more efficient and more reliable than traditional methods using instruments that need to be adjusted manually.

Many digitally controlled hearing aids have two or more memories. Thus, the hearing-aid user can change the electroacoustic characteristics of the hearing aid simply by activating one of its memories. For example, the electroacoustic characteristics that are best suited for a noisy acoustic environment can be stored in one memory, while the electroacoustic characteristics that are best suited for a quiet environment can be stored in another memory. The hearing-aid user can then choose the most appropriate set of electroacoustic characteristics for a given acoustic environment by activating the appropriate memory. This is usually done simply by pressing a button on the hearing aid.

Automatic adjustment of the frequency response of a hearing aid does not necessarily require the use of digital techniques, although these techniques provide considerably more flexibility in implementing automatic frequency-gain adjustment. It has been shown experimentally that the appropriate frequency response for a low-level signal is not the same as that for a high-level signal. Several modern hearing aids now provide automatic adjustment of the frequency response as a function of signal level. By this means it is possible for the hearing aid to adjust automatically to the appropriate frequency response for the signal being amplified.

An alternative method of signal processing that has recently been introduced in several modern hearing aids is that of multichannel amplitude compression. In hearing aids of this type, the signal to be amplified is split into two or more channels, and the gain in each channel is adjusted as a function of the input signal level in each channel. If, for example, background noise with strong, low-frequency components is present (a common occurrence), the gain in the low-frequency channel is reduced, thereby improving the overall quality of the amplified sound.

A problem with the use of digital techniques, at the present time, is that most of the electronic chips that have been developed for this purpose have been designed for applications other than hearing aids. As a consequence, most of the available chips for digital signal processing are too large and draw too much power for use in a practical hearing aid.

The electronic chips for a practical, wearable hearing aid (to be worn either behind the ear or in the ear canal) need to be extremely small and draw very little power from a small, low-voltage battery. As a result of these severe practical constraints, the use of digital techniques in modern hearing aids has progressed less rapidly than in other audio applications. Hybrid chips that allow for digital control of conventional analog (nondigital) circuits have already been developed, but chips for all-digital hearing aids have been developed only recently. Hearing aids that use digital techniques for processing the audio signal (rather than simply controlling conventional hearing aid amplifiers and filters) have only recently become available for clinical use.

The advantages of all-digital hearing aids are many. These include advanced methods of signal processing for noise reduction, speech enhancement, and feedback cancellation. All-digital master hearing aids for use in experimental studies have been developed by several research laboratories. These instruments have been of great value in studying ways in which digital signal-processing techniques could be of value in personal hearing aids. The results obtained thus far have been mixed. Some forms of signal processing have been found to be useful and are likely to be incorporated in the digital hearing aids of the future, while other advanced signal-processing techniques have not provided the anticipated improvements.

One of the most common complaints of hearing-aid users is that it is very difficult to understand speech in a noisy and/or reverberant acoustic environment. The problem of extracting speech from noise or reverberation is particularly difficult, and a satisfactory solution to the problem has yet to be found, even with the most advanced methods of signal processing. It is possible to remove background noise to some extent by using a frequency filter that adjusts automatically to the frequency content of the signals to be extracted from the noise. The philosophy underlying the design of these adaptive filters is to filter out those frequency components that are mostly noise, while leaving intact those frequency components that are mostly speech.

Experimental investigations of hearing aids with adaptive frequency filters have shown these instruments to be far from perfect. Important components of the speech signal are usually lost or distorted when the noise is removed. The net result is a reduction in the level of the background noise, thereby improving overall sound quality, but usually without any corresponding improvement in speech intelligibility. In many cases, there is a reduction in intelligibility. There are situations in which some improvement in speech intelligibility can be obtained with adaptive filtering, but only under special conditions that are not common, such as when the noise is very intense and consists of low-frequency components only.

The most promising approach to noise reduction with advanced signal processing techniques is to use more than one microphone. The differences in the signals from the various microphones are used to focus on sound coming from the direction of the speech signal and to discard sound (usually noise) coming from other directions. This technique works well when the speech and background noise are coming from different directions, which is often the case. It also works fairly well in a moderately reverberant room, but not in a highly reverberant environment; for example, in a room without carpets, curtains, or other sound absorbing surfaces.

An experimental hearing aid with multiple microphone inputs has already been developed, the additional microphones being mounted on the frame of a pair of eyeglasses. Under conditions typical of everyday use with speech and noise coming from different directions, the level of the background noise can be reduced by more than 6 dB (i.e., to less than a quarter of its power) with corresponding improvements in speech intelligibility. It remains to be seen whether an eyeglass hearing aid of this type will find favor among hearing-aid users. It should be noted that eyeglass hearing aids were fairly common in

Europe until they were replaced by smaller and more convenient behind-the-ear instruments.

The problem of enhancing speech intelligibility in quiet is less difficult than that of extracting speech from noise and/or reverberation. A substantial body of data has already been gathered on which sounds of speech are most difficult to understand for various hearing losses, and signal-processing techniques for enhancing these speech sounds are currently being developed. For example, weak voiceless consonants are often not heard by hearing-aid users. These sounds can be identified and enhanced automatically using modern signal-processing techniques. Experiments are currently in progress evaluating the extent to which this form of speech enhancement can improve intelligibility.

An important advantage of digital signal processing techniques is that they provide an extremely precise means for modifying both the amplitude and phase characteristics of the signal being amplified. A problem with many high-gain hearing aids is that of unstable acoustic feedback (i.e., the uncontrolled whistling that typically occurs when the gain is increased above a critical amount). It is possible to cancel unwanted acoustic feedback by inserting an electrical feedback signal that is equal in amplitude but opposite in phase to the acoustic feedback signal. This can be done without too much difficulty using digital signal-processing techniques, and a digital hearing aid with a significant amount of feedback suppression (the Danavox Genius) has been introduced recently. Hearing aids with more advanced methods of feedback suppression are currently being developed.

Whereas the requirements of small size and low power consumption impose major constraints on how much signal processing can be incorporated in practical behind-the-ear or in-the-canal hearing aids, these constraints do not apply to most assistive listening devices (ALDs), which are relatively large in size. As a consequence, ALDs with advanced signal-processing capabilities are a likely possibility for the near future. ALDs with superior sound quality have already been developed (e.g., infrared listening systems), as well as alerting systems for a variety of applications. ALDs for use with telephones are likely to show substantial changes in the future. At present, telephone ALDs are designed to serve primarily as amplification devices. Whereas improved amplification and improved methods of coupling hearing aids to telephone receivers will be developed as technology advances, a more important development will be the introduction of new types of telecommunication devices for use by deaf and hard of hearing people.

A service that was introduced recently and that has been of great value to deaf people is the telephone relay service. This service routes a telephone call from a hearing person to a typist at a relay center who types what is said on a text telephone. The output of the text telephone is transmitted to the deaf recipient of the telephone call. The deaf person can then respond either orally or by means of a text telephone, the received message being read by the typist to the hearing person.

The relay service, in its present form, has several major limitations.

1. Reduced speed and accuracy of communication, as compared to a typical spoken conversation between two hearing people
2. Lack of privacy resulting from the involvement of a third party (the typist)
3. The relatively high cost of operating a relay service that at present is labor intensive, a typist/operator being occupied full-time for the duration of each telephone call.

These problems can be addressed using modern signal-processing techniques.

Speed and accuracy of communication using a relay service can be improved substantially by using real-time stenography. A good stenographer can transcribe what is said with high accuracy and with a delay of no more than a second or two. The output of the stenograph, or any other device for shorthand typing, is fed into a computer that converts the shorthand symbols to legible text. The transcribed message is then transmitted to the deaf party by means of a text telephone in the usual way. This technique has been found to work extremely well in practice with telephone conversations proceeding at a rate comparable to that of normal voice communication. The use of real-time stenography is more expensive than conventional typing however, because of the higher rate of pay for professional stenographers. In addition to its possible application for improving the speed and relative accuracy of a telephone relay service, real-time stenography is being used increasingly for transcribing lectures at schools, universities, professional meetings, and business conferences involving deaf and hard of hearing people.

The second problem, lack of privacy using a telephone relay service, can be addressed by scrambling messages among several typists (or stenographers) so that no single typist transmits or receives an entire message, but rather deals with isolated phrases or sentences. This can be done when several relay calls are being handled simultaneously by a relay center. It is not an entirely satisfactory solution, but it is a step in the right direction.

The use of automatic speech recognition combined with text-to-speech synthesis would not only eliminate the privacy problem, but would also reduce costs substantially. The envisioned application of this technology would be for a relatively large, multiuser computer system to be placed at the relay center. A telephone call from a hearing person would be routed to the relay center where the speech is converted to text using the computer-based automatic speech recognition (ASR) system. The text would then be transmitted to the text telephone of the deaf person. The latter party could then respond orally, or by means of text that is converted to speech by a computer-based speech synthesizer located at the relay center.

ASR systems, at present, are not yet good enough to be incorporated into a practical telephone relay service. The primary limitations are accuracy of recognition, vocabulary size, speed of recognition, the need to train the computer for each speaker, and constraints on how the speech is produced; for example, whether the speech is produced continuously or one word at a time with a short pause between words (discrete word recognition). The best results are currently being obtained with discrete word recognizers. Typical values for a good quality discrete word ASR system, at the present time, are: vocabulary size in excess of 20,000 words, recognition accuracy greater than 95 percent for a cooperative speaker, and prior training of the ASR system for that speaker. The time taken for recognition to take place is on the order of one to several seconds. Similar results can be obtained with continuous speech, but for a much smaller vocabulary size. The thrust of current research in ASR is to make these systems speaker independent and to increase both accuracy of recognition and vocabulary size.

A discrete word recognizer of the above quality could be of practical value to a deaf or severely hard of hearing person in communicating with a small number of cooperative speakers who are willing to spend some time training the computer on their individual voices. Spouses, friends, and co-workers are likely to cooperate in this way, and the system could be of value in the home and office. Experimental evaluations of current generation ASR systems under these conditions are currently in progress. An additional advantage of the home or office application of ASR, is that the user can correct recognition errors made by the system as they occur. In a typical application of this type, the recognized speech appears on a screen and the user can make appropriate corrections by means of a keyboard. In the telephone relay application, the user does not have a screen for detecting recognition errors; hence, a greater degree of accuracy is required for ASR to be used in a practical telephone relay service.

It should be noted that ASR technology has already found a number of useful, albeit limited, applications for the general population, such as programming a VCR by voice. These applications are likely to grow in size and importance as ASR technology improves, thereby creating an increased demand for this technology and, in turn, reducing costs. ASR technology has much to offer deaf and hard of hearing people, and these developments should be watched closely. Perhaps the most important development of all is the growing integration of telephone, broadcast, and cable television technologies, with computer technology serving as the cement for binding these formerly independent methods of information transmission.

The telephone was originally designed to transmit speech by wire over long distances. From this simple beginning has grown a vast communications network allowing for interactive communication between almost any two points on this planet. This communication network is no longer limited to speech, but it is also used to transmit data, text, and visual images. The transmission of text and visual images is of particular value to deaf and hard of hearing people. These forms of information transmission include text telephones, electronic mail, computer notice boards, computer information services, and facsimile transmission (FAX).

The text telephone is the most widely used form of telephone communication among deaf people. Other, newly developed methods of text communication (e.g., electronic mail, FAX), however, have many advantages over traditional text telephones. These include a wider range of options (text, data, diagrams, photographs), as well as faster, more efficient information transmission at lower cost. Furthermore, these systems are widely used by the general population; hence, both their availability and the impetus for further technological advances are high.

At present, different devices are needed for text telephones, computer-based text and data transmission, and facsimile transmission. There is movement toward developing a single device for dealing with these different methods of information transmission. A personal computer with special plug-in boards can be used for this purpose, but some degree of computer literacy is required for its effective use. Easy-to-use, general-purpose communication devices are currently being developed that will automatically switch to the appropriate mode of communication (including voice transmission), depending on the signals received or transmitted.

Low-bandwidth video telephones are a new development that has yet to be shown to be economically viable. The currently available systems have been designed for use on conventional nondigital telephone lines and are not of a quality that is good enough for speechreading or sign language. An experimental video telephone designed specifically for use by deaf and hard of hearing people has been developed in England and is very promising. It is likely that when the proposed Integrated Services Digital Network (ISDN) becomes available within the next few years, the quality of video telephone transmission will be improved substantially, for the benefit of both normal-hearing and hearing-impaired telephone users.

The broadcast industry has focused on the use of wireless transmission for conveying information to a mass audience. Information was primarily one way, except for call-in programs in which members of the broadcast audience would place telephone calls to the radio or television station to express their views or to vote on a telephone poll. The latter activity is a recent development and is indicative of the growing trend for integrating different modes of information transmission.

The telephone industry has now moved into wireless transmission of information using cellular telephones, thereby combining the advantages of the vast telephone network with those of wireless transmission. For deaf and hard of hearing people, this development has both advantages and disadvantages. It is now possible, using a palm-top computer (i.e., a computer small enough to be held in the hand) coupled to a cellular telephone, for deaf and hard of hearing people to make more effective use of the telephone system.

Possible applications of cellular technology for deaf and hard of hearing people include:

1. Immediate reception of urgent text messages, similar to the use of a beeper by a physician
2. Communication between a conventional text telephone at a fixed location and a cellular-based text telephone at a roving location (e.g., on a ship, bus, train, airplane, or in some remote location)
3. Conversation with a hearing person using a cellular telephone and a telephone relay service
4. Face-to-face communication with a hearing person at any location, for example, in a shop or on the street using remote real-time stenography.

In application (4), the speech of the hearing person would be picked up by a microphone worn by the deaf (or severely hard of hearing) person and transmitted by cellular telephone to a stenographer for conversion to text that is then transmitted back to the display of a hand-held palm-top computer. The deaf person can then respond either orally, or by typing on the keyboard of the palm-top computer, which then converts the typed response into synthetic speech. An experimental trial of such a system has already been performed with positive results. The cost of operating the system, however, was relatively high because, in addition to the cost of a professional stenographer, two cellular telephone connections, one for the speech signal and one for the text, were required.

A disadvantage of cellular telephones for hard of hearing people is that the quality of the audio signal with present-day cellular telephones is poorer than that of conventional telephones. A hard of hearing person who has difficulty understanding speech using a conventional telephone will have even greater difficulty with a cellular telephone. These problems are likely to be resolved as the technology of cellular telephones improves. Work is currently in progress on developing digital cellular links that, among other applications, will carry voice communications with a reliability comparable to that of modern digital telephones.

A major limitation of the existing telephone network is that the wires conveying signals from the telephone exchange to individual telephones are limited in their capacity for conveying information. At best, a single video picture of moderate quality can be conveyed on existing telephone wires. In contrast, the wiring used for cable television can carry several hundred high-quality video channels. Until recently, legislation forced telephone and cable television companies to operate independently. This is no longer the case and with the change in legislation we are now seeing mergers and buyouts involving telephone and cable-television companies. As a consequence, telephone companies will have access to high-information-capacity cable connections to large numbers of homes and offices, while cable-television companies will have access to the vast telephone network and its switching capabilities.

The anticipated emphasis in the emerging "information superhighway" will be on interactive video communication and high-speed, low-cost data transmission. Both developments have important implications for assistive hearing technology. The cost of operating a text telephone, for example, should come down with revised pricing structures that take into account the amount of information transmitted as well as the connection time. Alternatively, signal transmission techniques could be developed that will transmit text messages in very short bursts of digital code so as to minimize the amount of time within which the telephone connection is in use.

The anticipated growth in the use of interactive video signal transmissions will provide new means of telecommunication for deaf and hard of hearing people. These will include high-definition video telephones at relatively low cost, individualized video shopping with captioning, electronic news magazines with both photographs and short video segments, interactive educational television, video conferences with captioning provided by real-time stenography, and various forms of entertainment such as movies on demand (with captioning), deaf theater, live telecasts, quiz shows with viewers as participants, interactive video games, and music television with specially processed sound tracks for hard of hearing viewers. Many other applications of interactive video technology are likely to emerge as people become acquainted with these techniques and their vast potential.

Modern technology can provide deaf and hard of hearing people with new possibilities for improving their quality of life in many ways. These opportunities depend on how this technology is used by both hearing and deaf people. The recently enacted Americans with Disabilities Act identifies the general direction in which modern technologies should be geared for this purpose, but it is up to users of modern technologies to explore their potential and to find out how best to take advantage of the opportunities presented.

The history of assistive hearing technology provides many examples of how new technologies developed for the general population have been of great practical value for both deaf and hard of hearing people. There have also been examples, however, where new technologies have placed deaf and hard of hearing people at a disadvantage. In order to take full advantage of the opportunities offered by emerging technologies and to minimize any concomitant disadvantages, consideration must be given to the development of practical means for accessing these technologies for the benefit of deaf and hard of hearing people.

Index